Harry Figi

Crystalline Organic Electro-Optic Microring Filters and Modulators

Harry Figi

Crystalline Organic Electro-Optic Microring Filters and Modulators

The first active organic crystalline microresonator marks an important step towards the use of organic crystals in integrated photonic devices

Südwestdeutscher Verlag für Hochschulschriften

Impressum/Imprint (nur für Deutschland/ only for Germany)
Bibliografische Information der Deutschen Nationalbibliothek: Die Deutsche Nationalbibliothek verzeichnet diese Publikation in der Deutschen Nationalbibliografie; detaillierte bibliografische Daten sind im Internet über http://dnb.d-nb.de abrufbar.
Alle in diesem Buch genannten Marken und Produktnamen unterliegen warenzeichen-, marken- oder patentrechtlichem Schutz bzw. sind Warenzeichen oder eingetragene Warenzeichen der jeweiligen Inhaber. Die Wiedergabe von Marken, Produktnamen, Gebrauchsnamen, Handelsnamen, Warenbezeichnungen u.s.w. in diesem Werk berechtigt auch ohne besondere Kennzeichnung nicht zu der Annahme, dass solche Namen im Sinne der Warenzeichen- und Markenschutzgesetzgebung als frei zu betrachten wären und daher von jedermann benutzt werden dürften.

Verlag: Südwestdeutscher Verlag für Hochschulschriften Aktiengesellschaft & Co. KG
Dudweiler Landstr. 99, 66123 Saarbrücken, Deutschland
Telefon +49 681 37 20 271-1, Telefax +49 681 37 20 271-0, Email: info@svh-verlag.de
Zugl.: Zürich, ETH, Diss., 2009

Herstellung in Deutschland:
Schaltungsdienst Lange o.H.G., Berlin
Books on Demand GmbH, Norderstedt
Reha GmbH, Saarbrücken
Amazon Distribution GmbH, Leipzig
ISBN: 978-3-8381-0999-2

Imprint (only for USA, GB)
Bibliographic information published by the Deutsche Nationalbibliothek: The Deutsche Nationalbibliothek lists this publication in the Deutsche Nationalbibliografie; detailed bibliographic data are available in the Internet at http://dnb.d-nb.de.
Any brand names and product names mentioned in this book are subject to trademark, brand or patent protection and are trademarks or registered trademarks of their respective holders. The use of brand names, product names, common names, trade names, product descriptions etc. even without a particular marking in this works is in no way to be construed to mean that such names may be regarded as unrestricted in respect of trademark and brand protection legislation and could thus be used by anyone.

Publisher:
Südwestdeutscher Verlag für Hochschulschriften Aktiengesellschaft & Co. KG
Dudweiler Landstr. 99, 66123 Saarbrücken, Germany
Phone +49 681 37 20 271-1, Fax +49 681 37 20 271-0, Email: info@svh-verlag.de

Copyright © 2009 by the author and Südwestdeutscher Verlag für Hochschulschriften Aktiengesellschaft & Co. KG and licensors
All rights reserved. Saarbrücken 2009

Printed in the U.S.A.
Printed in the U.K. by (see last page)
ISBN: 978-3-8381-0999-2

Contents

Abstract   3

Zusammenfassung   7

1 Introduction   11
- 1.1 Optical modulators in general 14
- 1.2 Linear electro-optic effect 17
  - 1.2.1 Nonlinear optics 17
  - 1.2.2 Frequency dispersion of the electro-optic effect 20
- 1.3 Electro-optic waveguide modulators 21
  - 1.3.1 Mach-Zehnder modulators 21
  - 1.3.2 Microring resonators 23
- 1.4 Advanced modulation formats 27
  - 1.4.1 High speed optical modulator components 27
  - 1.4.2 Non-return-to-zero and return-to-zero on-off keying 29
  - 1.4.3 IQ-modulator 30
  - 1.4.4 DBPSK 32
  - 1.4.5 DQPSK with a comparison to OOK and DBPSK 33
  - 1.4.6 PSK receiver 35
- 1.5 Summary 38
- 1.6 Outline of the thesis 39

2 Linear and nonlinear optical properties of DAPSH   43
- 2.1 Introduction 43
- 2.2 Crystal structure and sample preparation 44
- 2.3 Refractive indices 46
- 2.4 Absorption coefficients 47
- 2.5 Nonlinear optical properties 48
  - 2.5.1 Nonlinear optical tensor for second harmonic generation 48
  - 2.5.2 Maker fringe measurements 49
- 2.6 Conclusion 52

CONTENTS

3 Electro-optic single crystal waveguides of DAT2 — **53**
- 3.1 Introduction — 53
- 3.2 Material — 55
- 3.3 Sample preparation — 55
- 3.4 Crystal orientation and crystal quality — 58
- 3.5 Refractive indices and electro-optic modulation — 59
- 3.6 Growth of single crystalline nanowires and nanosheets — 62
- 3.7 Conclusion — 64

4 Electro-optic single-crystalline organic microring resonators — **67**
- 4.1 Introduction — 67
- 4.2 Material and crystal structure — 70
- 4.3 Sample preparation — 70
  - 4.3.1 Microring resonator fabrication — 70
  - 4.3.2 Crystal growth — 71
- 4.4 Transmission spectrum, resonance tuning and modulation — 73
- 4.5 Discussion and outlook — 77
- 4.6 Conclusion — 79

5 Conclusions and outlook — **81**
- 5.1 Conclusions — 81
- 5.2 Outlook — 83

A Appendix to chapter 2 — **85**
- A.1 Maker fringe measurements — 85
- A.2 Refractive index measurement — 86
- A.3 Refractive indices and coherence length — 87

B Appendix to chapter 3 — **89**
- B.1 Refractive index measurement — 89

C Appendix to chapter 4 — **95**
- C.1 DAT2 Mach-Zehnder modulators — 95

Bibliography — **101**

Acknowledgments — **113**

Abstract

Optical interconnects are presently the best choice for high-bandwidth data transmission as proven by the success of optical fiber communication in long-haul and local-area networks as well as in supercomputing rack-to-rack communication. In this context, where optical systems penetrate areas with smaller and smaller transmission distances, research and development efforts are now focusing onto even shorter interconnect distances: optical interconnections between intra-computer components at the board and even the chip level. As the demand for communication bandwidth in multi-core microprocessors rises, it is expected that metal interconnects will run into problems from excessive power consumption and latency in the foreseeable future and that optical interconnects offer better performance. Therefore, a research topic of high interest is to develop photonic integrated circuits, which could be used to establish very fast communication between circuit boards, between chips on a board, or even within single chips. To use light as information carrier on shortest distances, the development of highly integrated devices that are able to guide and confine light on a reduced size will be essential.

The optical waveguide-based modulator has emerged as a critical component to impose data on an optical signal and to control the flow of light signals in integrated high data rate communication systems. The modulator is sometimes specifically called external modulator to emphasize that it is separated from the light source. The growing importance of external modulators has taken place because this is the only known device that can modulate light signals at data rates of 100 Gbit/s and beyond, which are increasingly required data rates. To achieve the required level of very large scale integration, one possibility is to exploit the resonance characteristics of optical microcavities. Modulators based on ring-like optical cavities on micrometer scale are among the most promising components in optoelectronic integration.

In an external modulator, typically an electric voltage signal is used to control the properties of the light output from the modulator. Today, the fastest modulators used in long-haul transmission systems are electro-optic modulators based on lithium niobate ($LiNbO_3$). They exploit the electro-optic effect to change the refractive index of the optical $LiNbO_3$ medium by applying a voltage in order to modulate the phase of the optical carrier signal. Unfortunately, the velocity mismatch between the optical and electrical waves in $LiNbO_3$ limits the modulation

Abstract

speed to about 100 Gbit/s. On the other hand, electro-optic organic materials feature an electro-optic effect, which is of almost pure electronic origin and they have a very small velocity mismatch due to a low material dispersion of the dielectric constant. Therefore, organic materials are very suitable for high speed electro-optic modulators and indeed modulators based on organic materials with modulation frequencies greater than 100 GHz have been demonstrated.

To enable the tuning of the resonant properties of the rings by incorporating electro-optic materials is highly interesting, since the resonating structures can enhance the effect of refractive index changes on the transmission response. This explains the vast amount of studies devoted to the realization of electro-optic microring resonators in the last years, mainly focused on semiconductors, $LiNbO_3$ and polymers. Organic materials, with their beneficial ultra high-speed modulation response, can be divided into two main classes: poled polymers and organic electro-optic crystals. Polymers are potentially cheap and thin film processing is relatively easy, nevertheless they often show thermal as well as photochemical instabilities and require a challenging poling procedure to become electro-optically active. On the other hand, organic crystals have generally a superior thermal and photochemical stability, and can exhibit a highly polar order without poling, but growth of high quality thin films and waveguide structuring are still very challenging. Hence, it is of fundamental importance to develop new techniques for the fabrication of optical waveguides in organic crystalline materials.

The development of new waveguide structuring techniques of organic crystals has been the main focus of this thesis. Particularly, we have concentrated on techniques for melt-processable organic materials. Compared to the very common solution growth techniques, melt growth techniques exhibit relatively fast growth rates and do not possess the problem of solvent or solution inclusion into the crystal. Beside the demonstration of integrated organic electro-optic devices grown from the melt, also the latest results on the investigation of new organic nonlinear optical crystals with superior material properties compared to previously reported organic crystals are given. Such progress in the development of organic nonlinear optical materials with unprecedented nonlinear optical coefficients is extremely important for the realization of modulators with low operating voltages.

In an introductory chapter the essential advantages of incorporating organic materials with their ultra-fast electro-optic response into fundamental waveguide modulator designs are highlighted. The principles of basic electro-optic modulators are described. The introduced types of modulators (phase modulator, Mach-Zehnder modulator and microring resonator) have been realized in this work and are the fundamental building blocks to generate also advanced modulation formats, which have attracted considerable interest in recent years due to the ever increasing bandwidth demand.

In chapter 2, the properties of the organic material trans-4'-(dimethylamino)-N-phenyl-4-stilbazolium hexafluorophosphate (DAPSH), for optical components in high-speed communication systems, are investigated. DAPSH has been found to be a material with excellent nonlinear optical activity. The refractive index dispersion as well as the absorption spectra have been measured in bulk single crystals. The material features a large birefringence compared to other organic nonlinear optical materials of $\Delta n = 1.17 \pm 0.06$ at $\lambda = 0.83$ µm, due to the almost parallel packing of the nonlinear optical chromophores. The important nonlinear optical tensor elements have been determined by the Maker fringe technique at the fundamental wavelength of 1.9 µm. DAPSH crystals exhibit the presently largest crystalline non-resonant nonlinear optical susceptibility $\chi^{(2)} = 580 \pm 80$ pm/V.

In chapter 3, an integrated optical phase-modulator based on 2-(3-(2-(4-dimethylamino-phenyl)vinyl)-5,5-dimethylcyclohex-2-enylidene)malononitrile (DAT2) grown from the melt is demonstrated. The single-crystalline phase modulators were obtained with a new fabrication method, where the melt of the organic material flows into predefined channels by capillary force and crystallizes there upon cooling. For successful channel fabrication of the waveguide modulators, standard photolithography, plasma-etching and anodic bonding techniques have been combined to define channels, in which subsequently the crystals were grown. With this melt-based channel growth technique, the fabrication of single-crystalline optical waveguides with dimensions of about 1.5×4 µm^2 and their polar axis perpendicular to the waveguide channel was possible. Loss measurements have shown, that guiding losses were of about 14 dB/cm in the high refractive index contrast waveguides. Electro-optic phase modulation has been demonstrated, which allowed to determine the electro-optic coefficient r_{112} of DAT2 to be at least 7 ± 1 pm/V. Furthermore, the favorable growth characteristics of DAT2 combined with the melt-based channel growth technique allowed for the fabrication of single-crystalline nanowires and nanosheets inside large-area devices with crystal thicknesses below 30 nm and lengths of above 7 mm with very good optical quality.

The possibility of growing organic electro-optic single-crystals from the melt directly in pre-structured and electroded waveguide channels allowed subsequently also for the fabrication of integrated electro-optic Mach-Zehnder modulators and microring resonators presented in chapter 4. Electo-optic amplitude modulation in Mach-Zehnder interferometers has been demonstrated with DAT2 as the active material involved. The experimentally obtained half-wave voltage $\times$ length product $V_\pi \cdot L$ was determined to be 78 ± 2 Vcm for TE-modes and 60 ± 1 Vcm for TM-modes at a wavelength of 1.55 µm, which is in good agreement with the results obtained from the phase modulator in chapter 3. Furthermore, by using the melt-processable organic material 2-cyclo-octylamino-5-nitropyridine (COANP) the first electro-optically tunable organic single-crystalline microring resonators have been realized. Through the developed fabrication technique, the geometrical device parameters such as the height and width of both port and ring waveguide as well as the submicrometer gap between them could be accurately controlled, since only high quality inorganic materials had to be photolithographically processed. The rings had a radius of $R = 150$ µm, exhibited a quality Q-factor of $Q = 20'000$ and a high signal extinction ratio of about 10 dB. Their resonance spectrum could be electro-optically tuned at the rate of 0.13 GHz/V (1.1 pm/V) by applying an external electric field.

Abstract

We believe that the fabrication technique developed, together with the demonstration of the first active organic crystalline microresonators, is an important step towards the utilization of organic nonlinear optical active crystals in ultra-fast integrated devices for telecommunication applications. For a further miniaturization and efficiency increase of the waveguide structures toward very large scale integrated high-speed photonic devices, silicon/organic hybrid devices can be considered as promising route. Especially in the context, where the combination of organic electro-optic cladding materials with silicon-on-insulator nano-slot waveguides has been proposed, the fabricated nanowires appear to be extremely promising to allow for efficient filling of the ultra-small slot dimensions with an electro-optic material, avoiding the problems of poling and stability issues of polymers. The waveguide devices and nanowires based on single-crystalline organic materials are thus expected to be the starting point for a variety of applications in nonlinear photonics with the potential for high integration density and ultrahigh bandwidth operation of active optical modulators.

Zusammenfassung

Optische Kommunikationsverbindungen sind gegenwärtig die beste Möglichkeit um Datenübertragungen mit hoher Bandbreite zu realisieren, dies wird vom Erfolg von optischen Glasfaserkabeln, wie sie für Langreichweiten- und Lokal-Netzwerke sowie für Rack-zu-Rack Kommunikation in Supercomputern verwendet werden, eindrücklich untermauert. In Zusammenhang, wo optische Systeme in Anwendungsbereiche mit immer kürzerer Übertragungsdistanz vorstossen, zielen Forschungs- und Entwicklungsbemühungen darauf ab noch kürzere Verbindungsdistanzen zu realisieren: Optische Verbindungen zwischen computerinternen Komponenten auf der Karten- oder Chipebene. Mit den steigenden Anforderungen an die Kommunikationsbandbreite in Multikern-Mikroprozessoren ist davon auszugehen, dass metallische Verbindungen auf Grund übermässig grosser Leistungsaufnahme und Verzögerungszeit an ihre Grenzen stossen werden und dass optische Verbindungen ein höheres Leistungsvermögen aufweisen. Ein höchst interessanter Forschungsgegenstand ist deshalb die Entwicklung von hochintegrierten photonischen Schaltungen, die dazu verwendet werden können, eine sehr schnelle Kommunikation zwischen Computerkarten, zwischen verschiedenen Chips auf derselben Karte, oder sogar innerhalb einzelner Chips zu realisieren. Um Licht als Informationsträger auf kürzesten Distanzen verwenden zu können, ist die Entwicklung von Bauelementen erforderlich, welche es ermöglichen Licht auf kleiner räumlicher Grösse zu beschränken und zu lenken.

Der optische wellenleiterbasierte Modulator hat sich in den letzten Jahren zur kritischen Komponente entwickelt um Daten auf ein optisches Signal aufzumodulieren und die Signalführung in integrierten Hochgeschwindigkeits-Kommunikationsnetzwerken zu steuern. Der Modulator wird manchmal spezifisch externer Modulator genannt, um seine Abgetrenntheit von der Lichtquelle zu betonen. Die Wichtigkeit von externen Modulatoren hängt damit zusammen, dass sie die einzigen bekannten Bauelemente sind, die Lichtsignale mit Datenraten von 100 Gbit/s und mehr modulieren können, was zunehmend verlangte Datenraten sind. Eine Möglichkeit um das erforderliche Mass an Integrationsdichte für hochintegrierte Schaltungen zu erreichen, ist die Resonanzeigenschaften von optischen Mikrokavitäten auszunutzen. Modulatoren basierend auf mikrometergrossen ringähnlichen optischen Kavitäten sind mitunter die aussichtsreichsten Komponenten für die optoelektronische Integration.

Zusammenfassung

In einem externen Modulator wird typischerweise ein elektrisches Spannungssignal verwendet um die Eigenschaften des Lichts am Ausgang des Modulators zu steuern. Die heutzutage schnellsten Modulatoren, die in langreichweiten Netzwerken verwendet werden, sind elektrooptische Modulatoren basierend auf Lithiumniobat ($LiNbO_3$). Diese Modulatoren nutzen den elektro-optischen Effekt aus, um den Brechungsindex vom optischen Medium $LiNbO_3$ durch Anlegen einer Spannung zu verändern, um dabei wiederum die Phase des optischen Trägersignals zu modulieren. Unglücklicherweise wird die Modulationsgeschwindigkeit durch Phasenfehlanpassung zwischen der optischen und elektrischen Welle in $LiNbO_3$ auf etwa 100 Gbit/s gegrenzt. Andererseits zeigen elektro-optische organische Materialien einen elektro-optischen Effekt, welcher fast ganz elektronischen Ursprungs ist und sie haben eine sehr kleine Phasenfehlanpassung, da die dielektrische Konstante nur eine geringe Materialdispersion aufweist. Deshalb sind organische Materialien sehr gut für elektro-optische Hochgeschwindigkeitsmodulatoren geeignet, und in der Tat wurden bereits Modulatoren basierend auf organischen Materialien mit Modulationsfrequenzen von über 100 GHz realisiert.

Die Verwendung von elektro-optischen Materialen zur Beeinflussung der Resonanzeigenschaften der Ringe ist höchst interessant, da die Resonanzstruktur den Effekt der Brechungsindexänderung auf das Transmissionsverhalten verstärken kann. Dies erklärt die grosse Zahl der in den letzten Jahren durchgeführten Studien zur Realisierung von elektro-optischen Mikroresonatoren, wobei man sich hauptsächlich auf Halbleiter, $LiNbO_3$ und auf Polymere konzentrierte. Organische Materialen, die sich wie erwähnt durch ihre ultraschnelle Modulationsreaktion auszeichnen, können in zwei Hauptkategorien unterteilt werden: Gepolte Polymere und organisch elektro-optische Kristalle. Polymere sind potenziell günstiger und die Dünnfilmherstellung ist relativ einfach, nichtsdestoweniger zeigen Polymere häufig thermische sowie photochemische Instabilität und bedürfen eines anspruchsvollen Polungsschritts um elektro-optisch aktiv zu werden. Demgegenüber haben organische Kristalle eine im Allgemeinen bessere thermische und photochemische Stabilität und können eine äusserst polare Orientierung aufweisen ohne gepolt werden zu müssen, aber dafür ist das Wachstum von qualitativ hochwertigen Dünnfilmen und die Strukturierung von Wellenleitern sehr schwierig. Deshalb ist es eminent wichtig neue Techniken für die Herstellung von optischen Wellenleiterstrukturen in organisch kristallinen Materialen zu entwickeln.

Die Erarbeitung von neunen Strukturierungstechniken für Wellenleiter in organischen Kristallen war der Schwerpunkt dieser Forschungsarbeit. Insbesondere haben wir uns auf Techniken für aus der Schmelze verarbeitbare Materialien konzentriert. Verglichen mit der sehr gebräuchlichen Kristallzucht aus der Lösung, zeichnen sich Zuchttechniken aus der Schmelze durch eine schnellere Wachstumsrate aus und weisen nicht das Problem von Lösungsmitteleinschlüssen innerhalb der gewachsenen Kristalle auf. Neben der Demonstration von aus der Schmelze gewachsenen integrierten organischen elektro-optischen Bauelementen werden auch die neuesten Ergebnisse der Untersuchungen von neuen organischen nichtlinear optischen Kristallen präsentiert, welche bessere Materialeigenschaften zeigen als alle früher publizierten organischen Kristalle. Derartige Fortschritte in der Entwicklung von organischen nichtlinear optischen Materialien mit noch nie dagewesenen nichtlinear optischen Koeffizienten sind ausserordentlich wichtig für die Realisierung von Modulatoren mit niedrigen Betriebsspannungen.

In einem einführenden Kapitel werden die wesentlichen Vorteile der Verwendung von organischen Materialien mit ihrer ultraschnellen elektro-optischen Modulationsreaktion in grundlegenden Wellenleitermodulatoranordnungen hervorgehoben. Die Grundsätze von elementaren elektro-optischen Modulatoren werden geschildert. Die vorgestellten Modulatortypen (Phasenmodulator, Mach-Zehnder Modulator und Mikroresonator) sind im Rahmen dieser Arbeit realisiert worden und sind gleichzeitig die fundamentalen Bausteine um höherwertige Modulationsformate zu realisieren, welche in den letzten Jahren auf Grund des immer höheren Bandbreitenbedarfs an Bedeutung gewonnen haben.

In Kapitel 2 werden die Eigenschaften des organischen Materials trans-4'-(dimethylamino)-N-phenyl-4-stilbazolium hexafluorophosphate (DAPSH) für optische Komponenten in Hochgeschwindigkeits-Kommunikationssystemen untersucht. Es hat sich gezeigt, dass DAPSH ein Material mit herausragender nichtlinear optischer Aktivität ist. Die Dispersion der Brechungsindizes sowie das Absorptionsspektrum wurden in Einkristallen gemessen. Das Material zeichnet sich verglichen mit anderen organischen nichtlinearen Materialien durch eine hohe Doppelbrechung von $\Delta n = 1.17 \pm 0.06$ bei einer Wellenlänge $\lambda = 0.83$ µm aus, begründet durch die fast parallele Orientierung der nichtlinear optischen Chromophoren. Die wichtigsten nichtlinear optischen Tensorelemente wurden mit Hilfe der Maker fringe Methode bei einer fundamentalen Wellenlänge von 1.9 µm bestimmt. DAPSH Kristalle zeigen die gegenwärtig höchste kristalline nichtresonante nichtlinear optische Suszeptibilität $\chi^{(2)} = 580 \pm 80$ pm/V.

In Kapitel 3 wird ein integrierter optischer Phasenmodulator basierend auf dem aus der Schmelze gewachsenen Material 2-(3-(2-(4-dimethylaminophenyl)vinyl)-5,5-dimethylcyclohex-2-enylidene)malononitrile (DAT2) demonstriert. Der einkristalline Phasenmodulator wurde mit einer neuen Fabrikationsmethode hergestellt, in der die Schmelze des organischen Materials mittels Kapillarkraft in vordefinierte Kanäle fliesst und dort nach Abkühlen kristallisiert. Für eine erfolgreiche Wellenleitermodulatorherstellung wurden Standardphotolithographie, Plasmaätzprozesse und anodisches Bonden kombiniert, um Kanäle zu definieren, in welchen daraufhin die Kristalle gewachsen wurden. Mit dieser schmelzbasierten Kanalwachstumstechnik war es möglich einkristalline optische Wellenleiter mit Abmessungen von etwa 1.5×4 µm^2 und einer polaren Achse senkrecht zum Wellenleiter herzustellen. Verlustmessungen haben gezeigt, dass die Propagationsverluste der Wellenleiter, die einen hohem Brechungsindexkontrast aufweisen, bei etwa 14 dB/cm liegen. Elektro-optische Phasenmodulation konnte demonstriert werden, was es erlaubte den elektro-optischen Koeffizienten r_{112} von DAT2 auf einen Wert von mindestens 7 ± 1 pm/V festzulegen. Darüber hinaus ermöglichten die günstigen Wachstumseigenschaften von DAT2 kombiniert mit der schmelzbasierten Kanalwachstumstechnik die Herstellung von einkristallinen Nanofäden und Nanoschichten in grossflächigen Bauelementen

Zusammenfassung

mit Kristalldicken von unter 30 nm und Kristalllängen von über 7 mm mit sehr guter optischer Qualität.

Die Möglichkeit organisch elektro-optische Einkristalle aus der Schmelze in vordefinierten und mit Elektroden ausgestatteten Wellenleiterkanälen zu wachsen, ermöglichte schliesslich die Herstellung von integrierten elektro-optischen Mach-Zehnder Modulatoren und Mikroresonatoren, wie in Kapitel 4 dargestellt. Elektro-optische Amplitudenmodulation in Mach-Zehnder Interferometern wurde mit DAT2 als aktivem Material demonstriert. Das experimentell erhaltene Halbwellenspannung × Längenprodukt $V_\pi \cdot L$ wurde bei einer Wellenlänge von 1.55 µm für TE-Moden zu 78 ± 2 Vcm und für TM-Moden zu 60 ± 1 Vcm bestimmt, was sehr gut mit den Resultaten vom Phasenmodulator in Kapitel 3 übereinstimmt. Darüber hinaus wurde mit Hilfe des aus der Schmelze verarbeitbaren Materials 2-cyclo-octylamino-5-nitropyridine (COANP) der erste elektro-optisch aktive organische einkristalline Mikroresonator realisiert. Dank der entwickelten Fabrikationstechnik konnten die geometrischen Bauelementparameter wie die Höhe und Breite des Port- und Ringwellenleiters sowie der Submikrometerabstand zwischen diesen sehr präzise kontrolliert werden, da lediglich qualitativ hochwertige anorganische Materialien photolithographisch bearbeitet werden mussten. Die Ringe hatten einen Radius von $R = 150$ µm, wiesen einen Qualitätsfaktor von $Q = 20'000$ und ein hohes Extinktionsverhältnis von etwa 10 dB auf. Ihr Resonanzspektrum konnte durch Anlegen eines externen elektrischen Feldes mit einer Rate von 0.13 GHz/V (1.1 pm/V) elektro-optisch durchgestimmt werden.

Wir sind der Auffassung, dass die entwickelte Fabrikationsmethode, zusammen mit der ersten Demonstration eines aktiven organischen kristallinen Mikroresonators, ein wichtiger Schritt zur Verwendung von organischen nichtlinear optischen Kristallen in integrierten ultraschnellen Komponenten für Telekommunikationsanwendungen ist. Um eine weitere Miniaturisierung und Effizienzsteigerung der Wellenleiterstrukturen für deren Verwendung in hochintegrierten ultraschnellen photonischen Komponenten zu erreichen, werden silizium-organische Hybridbauelemente als sehr vielversprechend angesehen. Speziell in Zusammenhang, wo die Kombination von organisch elektro-optischen Cladding-Materialen mit "nano-slotted Silicon-on-insulator"Wellenleitern beabsichtig wird, erscheinen die hergestellten Nanoschichten als äusserst aussichtsreich, die nanometergrossen Furchen mit einem elektro-optischen Material effizient zu füllen und dabei die Probleme von Polymeren in Zusammen mit Polung und Stabilität zu umgehen. Wir gehen davon aus, dass die Wellenleiter und Nanoschichten basierend auf einkristallinen organischen Materialien den Ausgangspunkt für eine Vielzahl von Anwendungen in der nichtlinearen Photonik darstellen, mit dem Potential für hohe Integrationsdichte und Ultrabreitbandbetrieb von aktiven optischen Modulatoren.

CHAPTER 1

Introduction

In ancient times, the fastest method of long distance data transport was provided by smoke signals. There are other early demonstrations of free-space "optical" communication such as mirrors combined with sunlight used by the ancient Greeks or beacons invented in the 18th century in France, where semaphore arms were lifted and lowered, visible over long distances. In these early examples of "optical" communication the speed of data transport was acceptable, but capacity was low. In the middle of the nineteenth century Samuel Morse demonstrated the first relevant electrical communication circuit. This transition to electronic data transport was the basis to connect people around the world and to solve many problems related to transmission capacity, but about twenty years ago this technology began to feel its limits, while the need for an even higher communication capacity has continued to grow.

During the last 30 years, optical fiber transmission has played a key role in further increasing the bandwidth of telecommunication networks. The idea of using a light beam to carry an optical communication signal was not new. Alexander Graham Bell demonstrated the first free-space optical telephone message on June 3, 1880 using the patented "photophone" [1]. The photophone used sunlight as a source, modulated by reflecting off a vibrating mirror and photoconductive selenium as the receiver. This important invention is generally recognized as the forerunner of the modern fiber optical communication. However, Bell's idea had to wait about 90 years for better transmission systems and improved electronics to become useful in practical situations.

Todays fiber optic communication offers bandwidth and performance far superior to electrical cabling. But optical networks have been used for decades manly for larger transmission distances of at least of few kilometers. Nowadays, the ever-increasing amount of data traffic originating from an increasing number of high-bandwidth on-demand services, such as high definition TV or video conferencing, have shifted the need for ultrafast communications to a more local level. Therefore, there is a tendency for optical systems to penetrate areas with smaller and smaller transmission distances, many internet connections are already delivered to homes with optical

1. Introduction

fibers. The main advantage of the optical solution versus the electronic approach stems from the extremely high carrier frequency of the optical signal (~ 200 THz), which enables very high modulation frequencies (for instance, 100 GHz). For comparison, the augmented category 7 cable (CAT7A) is a highly advanced electrical cable standard for Ethernet over distances of 100 m and allows operation at frequencies of only 1 GHz. Therefore, optical communication is nowadays used for almost all high-speed (> 1 Gbit/s) interconnects running longer than 100 m, to overcome the bandwidth limitations imposed by electronic communication and to transmit more data at a faster rate.

Figure 1.1: MareNostrum was in November 2008 the eighth most powerful supercomputer in Europe [2] and uses ~ 5000 fiber links (orange cables) for rack-to-rack communications [4].

The gradual transition from electrical to optical interconnects for short-distance data communication applications is already clearly observable. For example, many of todays supercomputers use fiber-optic interconnects for rack-to-rack communications, such as the Roadrunner Petaflop supercomputer, the world's fastest computer in November 2008 [2], which is running at Los Alamos National Laboratory (Los Alamos, NM). An example of such fiber-optic interconnects, which appear as bundles of slim orange cables, is shown in Fig. 1.1. Especially for such high-performance computing, active optical cables have been introduced in the market as an alternative to copper links. As connectivity requirements reach or exceed 10 Gbit/s and the connection of more distant computers into clusters is needed, electrical interconnects experience severe bandwidth/distance limitations. Most of the active optical cables were developed for InfiniBand communication links, which are primarily used in high-performance computing and designed with copper ports. The form factor of the active optical cable transceiver mirrors the one of the copper cable it replaces and it acts therefore just like a copper-based cable. As their name implies, the active optical cables come with optical transceivers mounted on each end to provide electro-optical conversion and optical transmission. Since the fiber cable comes hard wired to the transceiver, most of the obstacles fibers usually encounter such as cleaning or splicing do not have to be considered. The fiber-optic cables are designed to transmit at data rates of up to 40 Gbit/s over distances of up to 300 m, depending upon the product under discussion. Intel, the first active optical cable supplier, introduced its Intel Connects Cables in Summer 2007 [3]. The cable provided four 5 Gbit/s lanes in each direction (20 Gbit/s unidirectional total bandwidth) and added only 550 ps of signaling latency per dual optical/electrical conversion. Other notable advantages of the optical cables are their size and weight. Because they contain no copper cores, the cables are 84% lighter than copper cables (see Fig. 1.2). For comparison, Luxtera has created the Blazar, an active optical cable that can support 300 m reach and enables 40 Gbit/s operation. However, the two introduced active optical cables use a fundamentally different approach for their transceivers. Luxtera's design uses a single optical light source split across four channels, where the light stream of every channel is modulated within a silicon chip and subsequently launched into the fiber. On the other hand, Intel Connects Cables use a vertical-cavity surface-emitting laser (VCSEL) array technology to launch light into the fiber. Table 1.1 compares the active optical cable products of two suppliers to an InfiniBand copper cable.

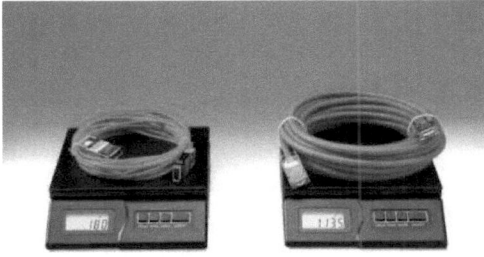

Figure 1.2: Intel Connects Cables [3] (left) are active optical cables, which are 84% lighter than copper equivalents (right).

Furthermore, computer processors presently use electrical signaling to receive and send data. With increasing processor complexity the number of input/output pads can not compete with the intrinsic microprocessor performance. The bandwidth per pin would have to be increased considerably, which is however challenging due to intrinsic bandwidth limitations of copper interconnects. Optical data communication over short-distances can provide a possible solution for the increasing data rates to and from a processor. Moreover, in the foreseeable future the electrical interconnects on the processor itself will show severe limitations in speed, power consumption, crosstalk and voltages needed to bias the devices [5], since future high-performance electronic integrated circuits of processors will count several billion transistors per chip and work with clock frequencies on the order of 10 GHz [6]. For example, the eight-core processor in Sony's Playstation 3 game console has a computation power of 256 billion floating point operations per second and it communicates with the peripheral graphics processor and memory at data rates of

Table 1.1: Active optical cables compared to conventional copper cables, where all products are primarily intended for high-speed technologies such as rack-to-rack interconnects in high-performance computing.

Type	Active optical cable		Copper cable
Product	Luxtera Blazar	Intel Connects Cables	InfiniBand
Active cable technology	Silicon Mach-Zehnder modulator	VCSEL array	-
Data rate/cable	40 Gbit/s	20 Gbit/s	40 Gbit/s
Maximum reach	300 m	100 m	30 m
Media weight/m/10 Gbit	2.5 g	5 g	30 g

25 Gbit/s, which is already within the range of concern for copper interconnection technology [7]. Optical interconnect technology is a promising alternative and is investigated by several research groups and industrial leaders. Optical interconnects can provide much greater bandwidth, lower power consumption, decreased interconnect delays, resistance to electromagnetic interference and reduced signal crosstalk [6]. Therefore, the replacement of metal by optical interconnects also at levels ranging down to chip-to-chip and intra-chip interconnections is highly desirable. The implementation of such optical interconnections relies critically on the development of micro-optical transceivers that are integrated with the microelectronics on chips. Thus, the development of solutions to efficiently perform switching and modulation of optical signals in a reduced size is very significant and could have a big impact on computer systems. In the following section 1.1 the todays most important optical modulators are reviewed and their individual advantages and disadvantages briefly discussed.

1.1 Optical modulators in general

Optical modulators used to encode electronic data signals onto a lightwave carrier are typically based on direct or external modulation. To generate an optical bit stream by direct modulation, an electrical signal is directly applied to a semiconductor laser used as an optical source. However, at high modulation rates, internal modulation likely introduces a change in the carrier frequency over time, which is referred to as chirp and highly undesirable especially in long-haul systems [8]. Therefore, chirping introduces limitations for direct modulation at high speeds, were on the other hand external modulation can be provided with negligible chirp.

The often applied external modulation techniques in integrated optics are usually based on the electro-absorption or the electro-optic effect. The mechanism of electro-absorption is based on the change of the absorption spectrum of the active material by applying an electric field. Evidentially the bandgap energy of the active material determines the wavelength of light that is absorbed. Via the so-called Franz-Keldysh effect the absorption can be increased for

1.1. Optical modulators in general

photon energies below the unbiased bandgap enegy by applying an electric field. To shift the wavelength for the optimal electro-absorption contrast to the telecom wavelength range around 1550 nm, tensile-strained GeSi can be used [9]. These SiGe modulators are compatible with the mature silicon technology since they can be processed using standard metal-oxide-semiconductor technology. However, since even the unbiased devices are operated very close to the absorption edge, losses are the dominant factor limiting their optical performance.

The other common approach to optical modulation is using electro-optic modulators. Certain materials change their refractive index when subjected to an electric field. The dependence of the refractive index change on the applied field can be either linear or quadratic. In the first case the effect is known as linear electro-optic effect or Pockels effect (see section 1.2), in the second case it is known as quadratic electro-optic effect or Kerr effect. In inorganic centrosymmetric crystals, the Kerr effect plays an important role near phase transitions, since the quadratic electro-optic coefficient is largely enhanced close to the Curie point. On the other hand, for the organic materials in our focus it is of minor importance. Hence, we concentrate on the discussion of the linear electro-optic effect.

The commonly used material to make electro-optic modulators based on the Pockels effect is the ferroelectric material lithium niobate ($LiNbO_3$). Standard integrated commercial optical modulators support data rates up to 40 Gbit/s (e.g. [10] or [11]). These devices based on an expensive bulk crystalline technology have been optimized to achieve very high modulation frequencies for the use with fiber-optics, but they perform signal processing with an unacceptably large size for on-chip integration (e.g. several centimeters for Avanex [11] and Covega [10] $LiNbO_3$ Mach-Zehnder modulators). To replace on-chip electrical with optical interconnects, a high level of integration is required. Especially silicon has attracted a lot of interest for such a high density integration, since its high refractive index and low absorption in the telecommunication wavelength range around 1.55 µm allow for excellent optical mode confinement and guiding. Also the matured processes, the access of a huge knowledge base and the ability to pattern nanoscale structures makes it an attractive material system for integrated photonics. The implementation of optics in silicon substrates can also profit from the choice of silicon-on-insulator substrates for mainstream high-performance electronic circuits. Especially the mentioned good high-index guiding properties of silicon-on-insulator devices have led to several impressive results considering the miniaturization of photonic circuits [12, 13].

On the other hand, the Pockels effect is not observed in silicon due to the centro-symmetric nature of the crystal structure. However, there is another important technique to affect the silicon refractive index, which is based on carrier injection or depletion. Since the modulation speed for this free-carrier dispersion effect is determined by the rate at which carriers can be injected and removed, the long recombination carrier lifetime in the silicon region generally keeps the bandwidth of such silicon-based devices within the GHz region (e.g. the fastest free-carrier injection broadband light modulation in silicon technology allowed modulation up to 30 GHz [14]). To achieve higher modulation bandwidth, the device capacitance has to be reduced for example by reducing the silicon resistance with higher doping. This approach, however, translates to higher transmission losses because losses increase with increased doping. In addition such charge injection devices typically require a high driving power for obtaining a significant modulation depth.

In addition to ferroelectric crystals like $LiNbO_3$, there are also special organic materials, which exhibit a strong electro-optic effect. In organic materials the linear electro-optic response is mainly determined by the polarizability of the molecules, which arises from π electrons between a donor and an acceptor group, whereas intermolecular interactions are of minor importance. Therefore, the electro-optic effect in organic materials is mainly of electronic origin and almost dispersion free. On the other hand in inorganic materials, which are based on a strong bonding between the lattice components (ions), lattice vibrations play a dominant role at low frequencies, since the ions act as additional polarizable elements. Organic materials feature in general not only a lower dispersion of the electro-optic coefficient but also of the dielectric constant compared to inorganic materials. Therefore, for organic devices velocity matching between the optical and microwave in traveling-wave modulators is easier to realize and a much smaller portion of the applied voltage drops over the buffer layer of a waveguide. For example, high-speed electro-optic modulators based on organic polymers with high bandwidth of up to 165 GHz and extremely low switching voltages below 1 V have been recently demonstrated [15–18]. These impressive results show the potential of organic materials in future high-speed modulators, where contrariwise there are strong doubts about the feasibility of $LiNbO_3$ based electro-optic Mach-Zehnder modulators with a bandwidth above 100 GHz [19, 20].

The most common attempt to quantify the quality of a modulation effect is to measure the refractive index shift obtained for a given modulation voltage. In this context the most important property of a modulator is the half-wave voltage V_π. This is the voltage required for inducing a phase change of π, which corresponds to go from the operation point with minimum transmission to that with maximum transmission. Today's best conventional all-polymer modulators feature a halfwave voltage around $V_\pi = 2$ V to modulate light with a wavelength around 1.55 µm in $L = 1$ cm long Mach-Zehnder modulators [16, 18, 21]. Usually the length-independent product $V_\pi \cdot L$ is reported, where a lower value is desirable, because it reduces the amount of power needed to switch the modulator.

For completeness it shall be noted that slotted silicon waveguide configurations, which combine the highly nonlinear characteristics and the ultra-fast electro-optic effect of organic materials with the good high-index guiding properties of silicon-on-insulator devices [22, 23], could be advantageous for electro-optic modulators. Slot waveguide modulators allow for a considerable improvement of the tunability over the current state of the art [22, 24]. The essential feature of a slot waveguide is that two high-index silicon stripes are separated by a small distance on the order of 20-140 nm filled with a lower-index material. The optical field intensity in such a structure tends to concentrate within the low-index slot for the polarization perpendicular to the slot. Therefore and also due to the close proximity of the electrical contacts in a slotted geometry filled with an electro-optic active material, a very large refractive index change for a given modulation voltage can be obtained compared to a more conventional waveguide with external electrodes. In a Mach-Zehnder configuration $V_\pi \cdot L$ values of around 0.5 Vcm have recently been obtained [25] and theoretical studies suggest that it may be possible to obtain slot-waveguide-based silicon/organic hybrid modulators with $V_\pi \cdot L$ values of 4 mVcm, an improvement over currently best $V_\pi \cdot L$ values by about 2 orders of magnitude [24].

In the following section 1.2 a description of the electro-optic effect, which is of special interest concerning the realization of integrated optical modulators in organic materials, is given.

1.2 Linear electro-optic effect

Linear electro-optic phenomena only appear in materials that feature a second-order nonlinear optical response. Therefore, they can be studied in the general framework of nonlinear optics. In this section a short introduction to nonlinear optics is given and at the end the linear electro-optic effect is on one hand derived from general nonlinear effects and on the other hand the usual definition based on the deformation and rotation of the optical indicatrix due to an applied electric field is given.

1.2.1 Nonlinear optics

The following discussion of electro-magnetic waves is restricted to materials, where free charges and currents are absent. The Maxwell equations in SI units in such a medium are given by [26]

$$\nabla \times \boldsymbol{E} = -\frac{\partial \boldsymbol{B}}{\partial t}, \tag{1.1}$$

$$\nabla \times \boldsymbol{H} = \frac{\partial \boldsymbol{D}}{\partial t}, \tag{1.2}$$

$$\nabla \cdot \boldsymbol{D} = 0 \tag{1.3}$$

$$\nabla \cdot \boldsymbol{B} = 0, \tag{1.4}$$

where $\boldsymbol{E}$ and $\boldsymbol{H}$ are the electric and magnetic fields, $\boldsymbol{B} = \mu_0 \boldsymbol{H}$ is the magnetic induction in a non-magnetic medium, $\boldsymbol{D}$ is the dielectric displacement and μ_0 is the vacuum permeability. The electric field, present in a dielectric material, will lead to a redistribution of the charges, hence resulting in an induced polarization $\boldsymbol{P}$. The constitutive relation between the dielectric displacement and the electric field is given by [26]

$$\boldsymbol{D} = \varepsilon_0 \boldsymbol{E} + \boldsymbol{P}, \tag{1.5}$$

where ε_0 is the dielectric permittivity in vacuum. Applying the operator $\nabla \times$ to both sides of Eq. (1.1) and using Eqs. (1.2) and (1.5) one obtains

$$\nabla \times \nabla \times \boldsymbol{E} + \frac{1}{c^2}\frac{\partial^2 \boldsymbol{E}}{\partial t^2} = -\mu_0 \frac{\partial^2 \boldsymbol{P}}{\partial t^2}, \tag{1.6}$$

1. Introduction

where $c = 1/\sqrt{\varepsilon_0 \mu_0}$ denotes the speed of light in vacuum. To solve this equation, the induced polarization has to be expressed as a function of the electric field, which is normally rather complicated. Nevertheless, assuming that the material is reacting instantaneous to the applied electric field, the polarization can be expanded in a Taylor series (in Ref. [27] a more general theory, which considers a non-instantaneous relation between the time-dependent vectors $\boldsymbol{E}(t)$ and $\boldsymbol{P}(t)$ is discussed). Using Einstein's summation convention, the induced polarization is given by

$$P_i = \varepsilon_0 \chi_{ij}^{(1)} E_j + \varepsilon_0 \chi_{ijk}^{(2)} E_j E_k + \varepsilon_0 \chi_{ijkl}^{(3)} E_j E_k E_l + \ldots, \tag{1.7}$$

where $\boldsymbol{\chi}^{(n)}$ is the nth-order susceptibility tensor (in literature $\boldsymbol{\chi}^{(n)}$ is sometimes referred to as the $(n^{\text{th}} - 1)$-order nonlinear susceptibility tensor). The denominators of the Taylor expansion are included in the definition of the susceptibilities. For symmetry reasons, the odd-order susceptibilities are present in any material, whereas the even-order ones only occur in non-centrosymmetric materials. Splitting up the induced polarization term in a linear and a nonlinear term

$$P_i = P_{\text{L},i} + P_{\text{NL},i} \tag{1.8}$$

with

$$P_{\text{L},i} = \varepsilon_0 \chi_{ij}^{(1)} E_j \tag{1.9}$$

$$P_{\text{NL},i} = \varepsilon_0 \chi_{ijk}^{(2)} E_j E_k + \varepsilon_0 \chi_{ijkl}^{(3)} E_j E_k E_l + \ldots, \tag{1.10}$$

the wave equation (1.6) can be simplified and we obtain

$$\nabla \times \nabla \times \boldsymbol{E} + \frac{\boldsymbol{\varepsilon}}{c^2} \frac{\partial \boldsymbol{E}}{\partial t^2} = -\mu_0 \frac{\partial \boldsymbol{P}_{\text{NL}}}{\partial t^2}, \tag{1.11}$$

where $\boldsymbol{\varepsilon} = 1 + \boldsymbol{\chi}^{(1)}$ is the dielectric tensor. The nonlinear polarization $\boldsymbol{P}_{\text{NL}}$ acts as a source term in the equation. Note that for crystals with anisotropic dielectric properties $\nabla \cdot \boldsymbol{E}$ is generally not equal to zero and hence the wave equation can not be further simplified.

Nonlinear optical effects

In the following the optical properties of a nonlinear medium are examined, in which nonlinearities of order higher than the second are negligible. It is convenient to relate the Fourier component of the induced nonlinear polarizability to the one of the present electric fields. To do so, the time-dependent electric field $\boldsymbol{E}(\boldsymbol{r},t)$ as well as the nonlinear optical polarization $\boldsymbol{P}_{\text{NL}}(\boldsymbol{r},t)$ have to be written as a superposition of monochromatic fields $\boldsymbol{E}^{\omega_p}(\boldsymbol{r})$ and $\boldsymbol{P}^{\omega'_p}(\boldsymbol{r})$ with distinct frequencies ω_p and ω'_p:

$$\boldsymbol{E}(\boldsymbol{r},t) = \frac{1}{2} \sum_p \left(\boldsymbol{E}^{\omega_p}(\boldsymbol{r}) e^{-i\omega_p t} + \text{c.c.} \right) \tag{1.12}$$

$$\boldsymbol{P}_{\text{NL}}(\boldsymbol{r},t) = \frac{1}{2} \sum_p \left(\boldsymbol{P}^{\omega'_p}(\boldsymbol{r}) e^{-i\omega'_p t} + \text{c.c.} \right). \tag{1.13}$$

1.2. Linear electro-optic effect

If two monochromatic electric waves with frequency ω_1 and ω_2 are present in the medium, a multitude of nonlinear polarizations oscillating with different frequencies are obtained. For example, the expansion gives in the case of sum-frequency generation for the selected frequency $\omega_3 = \omega_1 + \omega_2$

$$P_{\text{NL},i}^{\omega_3} = \varepsilon_0 \chi_{ijk}^{(2)}(-\omega_3, \omega_1, \omega_2) E_j^{\omega_1} E_k^{\omega_2}. \tag{1.14}$$

Second-harmonic generation (SHG) is a special case of sum-frequency generation with one incident wave with frequency $\omega = \omega_1 = \omega_2$. The nonlinear polarization for SHG is usually expressed by the nonlinear optical coefficient d_{ijk}, often used for the characterization of nonlinear optical materials (see also Appendix A.1):

$$P_{\text{NL},i}^{2\omega} = \frac{1}{2}\varepsilon_0 \chi_{ijk}^{(2)}(-2\omega, \omega, \omega) E_j^{\omega} E_k^{\omega} = \varepsilon_0 d_{ijk}(-2\omega, \omega, \omega) E_j^{\omega} E_k^{\omega}. \tag{1.15}$$

Analog to the case of sum-frequency generation, difference-frequency generation is the result of mixing two input beams to produce an output beam of frequency $\omega_2 = \omega_3 - \omega_1$. The nonlinear polarization of this process can be written as

$$P_{\text{NL},i}^{\omega_2} = \varepsilon_0 \chi_{ijk}^{(2)}(-\omega_2, \omega_3, -\omega_1) E_j^{\omega_3} (E_k^{\omega_1})^*. \tag{1.16}$$

The linear electro-optic effect can be derived as well in the framework of second order nonlinear effects. For the electro-optic effect one of the interacting fields $\boldsymbol{E}^{\Omega} = \frac{1}{2}\left(\boldsymbol{E}_0^{\Omega} \cdot e^{-i\Omega t} + \text{c.c.}\right) \approx \boldsymbol{E}_0^0$ is quasi-static and the other field $\boldsymbol{E}^{\omega} = \frac{1}{2}\left(\boldsymbol{E}_0^{\omega} \cdot e^{-i\omega t} + \text{c.c.}\right)$ is at an optical frequency ω, where $\omega \gg \Omega \approx 0$. The contribution to the sum- and difference-frequency generation in the induced polarization is then given by (only terms proportional to $\boldsymbol{E}^{\omega}$):

$$P_{\text{NL},i}^{\omega} = 2\varepsilon_0 \chi_{ijk}^{(2)}(-\omega, \omega, 0) E_j^{\omega} E_k^{0}. \tag{1.17}$$

Equation (1.17) provides a linear relation between $\boldsymbol{P}_{\text{NL}}^{\omega}$ and $\boldsymbol{E}^{\omega}$, which can be written in the form $\boldsymbol{P}_{\text{NL}}^{\omega} = \varepsilon_0 \Delta \boldsymbol{\chi} \boldsymbol{E}^{\omega}$, where $(\Delta \chi)_{ij} = 2\chi_{ijk}^{(2)} E_k^0$ represents a change in the susceptibility proportional to the amplitude of the electric field $\boldsymbol{E}^{\Omega}$. The corresponding change Δn of the refractive index n depends as well linearly on the quasi-static electric field $\boldsymbol{E}^{\Omega}$, which can be seen for example by differentiating the relation $n^2 = 1 + \chi$.

Linear electro-optic effect

The electro-optic effect is generally not described as a nonlinear optical effect, but rather defined as the deformation and rotation of the optical indicatrix, which is defined as

$$\left(\frac{1}{n^2}\right)_{ij} x_i x_j = 1 \tag{1.18}$$

in a Cartesian coordinate system (x_1, x_2, x_3). An electric field can now change the coefficients according to

$$\Delta\left(\frac{1}{n^2}\right)_{ij} = r_{ijk} E_k, \tag{1.19}$$

19

1. Introduction

where r_{ijk} is the electro-optic tensor, which is symmetric in the first two indices. In most practical cases $\Delta n \ll n$, and therefore Eq. (1.19) can be written as a linear change of the refractive index. If this change occurs along the main axes of the optical indicatrix, it can be written as

$$\Delta n_i \approx -\frac{1}{2} n_i^3 r_{iik} E_k. \qquad (1.20)$$

To attain large refractive index changes, the quantity $n_i^3 r_{iik}$, which is defined as figure of merit for electro-optic modulators, has to be maximized.

3rd order nonlinear effects

3rd order nonlinear optical effects involving $\chi^{(3)}$ are the quadratic electro-optic effect $\Delta n \propto E^2$ (Kerr effect), the optical Kerr effect, which describes a light-intensity induced change in refractive index $\Delta n \propto I$, third harmonic generation, general four-wave mixing and others. An overview of most known nonlinear optical effects can be found for example in [27].

1.2.2 Frequency dispersion of the electro-optic effect

The coefficients r_{ijk} of the electro-optic effect depend on the modulation frequency Ω of the applied electric field as well as on the light frequency ω. The dependence on ω is associated with the optical dispersion of the material, which relates to its electronic response. It is determined by the frequency and strength of the active electronic transitions. On the other hand, the dependence on Ω arises from the additional lattice contribution to the electro-optic effect. The free electro-optic effect r^{T} has an electronic part r^{e} and contributions from optic r^{o} and acoustic r^{a} phonons [28]

$$r^{\text{T}} = r^{\text{a}} + \underbrace{r^{\text{o}} + r^{\text{e}}}_{r^{\text{S}}}. \qquad (1.21)$$

The clamped electro-optic coefficient r^{S} arises from the fact that in experiments where the crystal is mechanically clamped, it cannot be deformed and the piezo-electric contributions r^{a} of the acoustic phonons are suppressed. For non-covalently bound organic crystals, the lattice contributions are weak and r is dominated by the electronic charge transfer between donor and acceptor. For inorganic compounds $r(\Omega)$ shows much stronger resonance peaks at frequencies below 10^{13} Hz, since the electro-optic coefficient has large contributions from lattice vibrations (see Fig. 1.3). A flat frequency response up to the maximum modulator frequency is preferred, as it is generally found in organic materials. A comparison of the electro-optic coefficients and its contributions is given in Table 1.2 for the widely known inorganic crystals $LiNbO_3$ and potassium niobate ($KNbO_3$) and the organic crystal DAST (4-N, N-dimethylamino-4'-N'-methyl-stilbazolium tosylate).

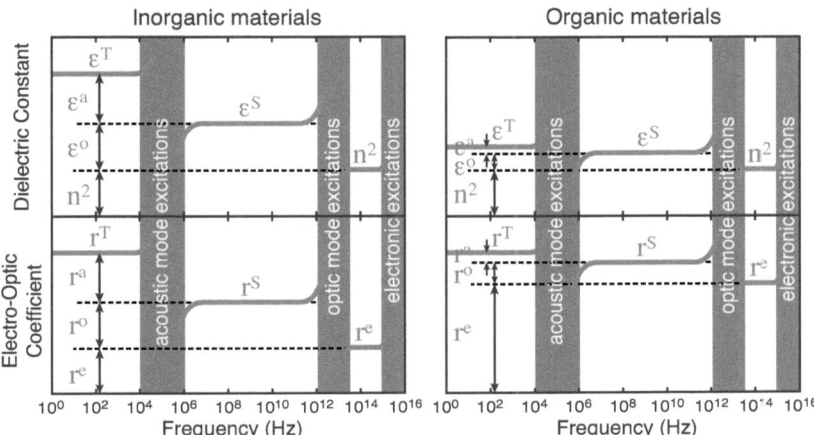

Figure 1.3: Simplified dispersion of the dielectric constant and the electro-optic coefficient as a function of the modulation frequency Ω of the applied electric field for organic and inorganic materials [28].

Table 1.2: Comparison of the electro-optic properties of the inorganic crystals LiNbO$_3$ and KNbO$_3$ with the organic crystal DAST. Given are the static dielectric constant ε_{r}, the squared refractive index n^2, the free and clamped electro-optic coefficient and its electronic contribution, as well as the nonlinear optical coefficient d. The values for LiNbO$_3$ and KNbO$_3$ are specified for the optical wavelength $\lambda = 632.8$ nm and for DAST at 1.5 µm [28].

	ε_{r}	n^2	r^{T} [pm/V]	r^{S} [pm/V]	r^{e} [pm/V]	d [pm/V]
LiNbO$_3$	29	4.5	30.8	30.8	5.8	36
KNbO$_3$	44	4.5	63	34	5.0	27.4
DAST	5.2	4.6	47	48	36	290

1.3 Electro-optic waveguide modulators

The realization of organic electro-optic crystalline Mach-Zehnder modulators as well as microring resonators is demonstrated in chapter 4. In this section a short theoretical description of these integrated photonic devices is given.

1.3.1 Mach-Zehnder modulators

A widely used modulator type is the Mach-Zehnder modulator, which is schematically drawn in Fig. 1.4. The operation principle of such a device is discussed in the following. The intensity I_{in} of the optical carrier is divided at the first Y-junction of the interferometer. Before recombining

1. Introduction

at the second Y-junction, the optical waves in the two arms experience an opposite phase shift, controlled by the voltage V_{in}, applied to the electro-optically active waveguide segments. If no voltage is applied, the waves arrive at the output in phase and the total power is combined. The OFF state is obtained by introducing a phase shift of π between the separated fields. This antisymmetrical wave cannot exit the output and is being reflected or radiated into the surrounding material. If the in-coupled beam has been divided into two equal parts, the transmitted optical intensity is described by the relation

$$\frac{I_{out}}{I_{in}} = \cos^2\left(\frac{\Delta\phi}{2}\right) = \cos^2\left(\frac{\pi V_{in}}{2 V_\pi}\right). \tag{1.22}$$

The phase shift between the two arms of the interferometer with length L is denoted by $\Delta\phi$ and the half-wave voltage V_π is the voltage required to produce a phase shift of π at the end of the two interferometer arms. The switching curve (1.22) is shown in Fig. 1.4.

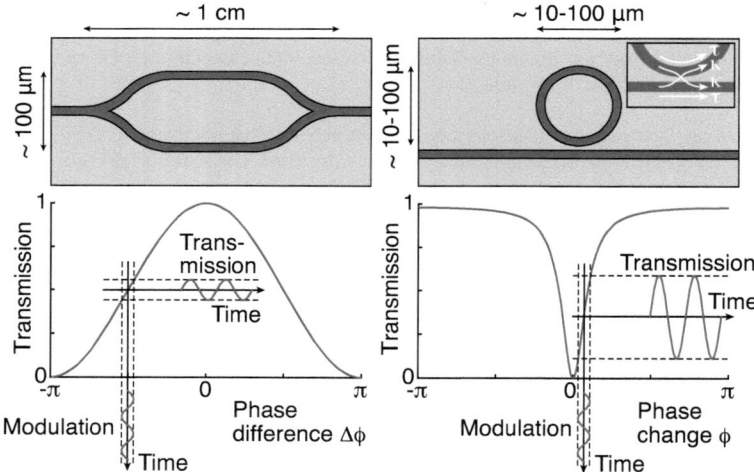

Figure 1.4: Electro-optic modulation using a Mach-Zehnder modulator (left) and a microring resonator (right). The transmission function for the Mach-Zehnder modulator is plotted according to Eq. (1.22) and for the microresonator according to Eq. (1.31). Indicated is as well the change in transmission induced by applying a sinusoidal modulation voltage. The inset in the schematic drawing of the microring resonator shows the coupling region enlarged; κ and τ are the electric field coupling and transmission coefficients, respectively.

The phase of the optical wave in one arm of the interferometer is controlled by the electro-optic effect. In a homogenous electric field, the refractive index change induced in the electro-optic material is given by Eq. (1.20). For general waveguides, the effective index n_{eff} of a guided mode is a function of the refractive index n of the electro-optic material involved in the waveguide structure. Since the electro-optic change of refractive index is usually less than 1 ‰,

1.3. Electro-optic waveguide modulators

the modal effective index change can be approximated with high precision as

$$\Delta n_{\text{eff}} = \left.\frac{\partial n_{\text{eff}}}{\partial n}\right|_n \cdot \Delta n = \xi_n \cdot \Delta n, \qquad (1.23)$$

where Δn is the electro-optically induced refractive index change. However a slightly altered form of Eq. (1.20) has to be used to calculate Δn due to the inhomogenous electrical and optical fields in real modulator devices [29]

$$\Delta n = -\frac{n^3 r}{2}\frac{\Gamma}{G}V_{\text{in}}, \qquad (1.24)$$

where G is the gap between the electrodes, V_{in} is the applied voltage to the electrodes and r is the electro-optic coefficient of the active material. Γ represents the spatial overlap of the optical intensity profile with the applied electric field and is given by [30]

$$\Gamma = \frac{G}{V_0}\frac{\iint E_o^2(x,z)E_m(x,z)\,\mathrm{d}x\,\mathrm{d}z}{\iint E_o^2(x,z)\,\mathrm{d}x\,\mathrm{d}z}, \qquad (1.25)$$

where E_o is the optical field and E_m the field strength of the modulation field in the material, V_0 is the amplitude of the applied modulation voltage $V_{\text{in}} = V_0 e^{i\Omega_0 t}$ and propagation of the optical mode in y-direction has been chosen. The integral should only encompass over the electro-optic active area ($r \neq 0$) of one arm.

Therefore, the phase shift between the two arms in presence of a modulation voltage V_{in} is given by

$$\Delta\phi = 2k_o\Delta n_{\text{eff}}L = -2k_o\xi_n \cdot \frac{n^3 r}{2}\frac{\Gamma}{G}LV_{\text{in}}, \qquad (1.26)$$

where k_o is the wave number of the optical mode in vacuum. The half-wave voltage V_π is given by

$$V_\pi = \frac{\lambda}{2\xi_n}\frac{1}{n^3 r}\frac{G}{\Gamma L}, \qquad (1.27)$$

where λ is the wavelength in vacuum of the optical field. Since low half-wave voltages are desirable, a high electro-optic figure of merit $n^3 r$ and a large overlap Γ between optical and modulation field are beneficial.

1.3.2 Microring resonators

In this section, a brief introduction to the topic of microring resonators is given. Their performance with respect to transmission function and switching voltage is described and compared to those of Mach-Zehnder modulators. A detailed description can be found in standard textbooks (e.g. [31, 32])

To achieve the required level of chip-scale optical integration, two classes of devices are currently emerging as a promising route: On one hand microring resonators and on the other hand photonic bandgap structures, where both concepts may additionally include the combination of passive high index waveguides with electro-optic active materials similar to the slotted silicon waveguides mentioned in section 1.1 [22, 24, 33]. Currently, an important

1. Introduction

criterion in favor of microring resonators is the more sophisticated fabrication requirements set by photonic bandgap structures, which makes it more difficult to realize low-cost devices. Microring resonator devices are based on a resonant light-confining structure that enhances the sensitivity of light to small changes in refractive index and enables high-speed operation. Commonly, they consist of an ordinary waveguide that channels light in a closed loop. In general the loop can take different forms of closed shapes, such as a circular ring, a racetrack or an ellipse. Coupling between the resonant cavity and the ordinary waveguide is achieved through evanescent coupling, which in turn requires to bring the two components in close proximity. Component wavelengths of an optical signal coupled into the closed waveguide configuration are resonant with the cavity if its effective circumference supports an integer number of wavelength. Thus, at their must fundamental level, microring resonators act as spectral filters, due to their strong wavelength sensitivity, which makes them attractive for the processing of optical signals.

Ring resonators made of electro-optic active materials are of special interest, since the optical ring length can be altered by the electro-optic effect and thus the resonance wavelength tuned. Active microring resonators can be therefore used as switchable wavelength filters and also as electro-optic modulators. Active microring devices have been demonstrated up to now in inorganic crystals, e.g. $LiNbO_3$ [34] and electro-optic poled polymer devices [35–37]. Electrical modulation based on carrier injection has also been demonstrated in silicon microrings [12, 38].

Basic principle of microring resonators

A microring structure consisting of a straight and a ring waveguide in close proximity is schematically drawn in Fig. 1.4. As mentioned above, if the optical ring length L is a multiple of the optical wavelength, then the field in the ring is adding up constructively and therefore a dip in the transmission curve is observed. The transmission as a function of the round trip phase delay $\phi = k_0 n_{\text{eff}} L$ depicted in Fig. 1.4, shows similar features as the one of a Fabry Perot resonator, where, however, the feedback is achieved by back reflection instead of coupling. As often in literature, the real part $n_{\text{eff}} = \text{Re}(N_{\text{eff}})$ of the complex effective index N_{eff} is simply denoted as effective index here.

The important quantities, which describe a Fabry Perot resonator as well as a microring resonator, are the resonator losses and the strength of the feedback. In a microring resonator the intrinsic losses stem from radiation to the substrate and cladding, because of the bend shape of the waveguide structures, while additional losses occur due to fabrication imperfections and material absorption. The decay constant of the field inside the cavity after one round trip is given by $\sigma = e^{-\frac{\alpha}{2}L}$ with $\alpha = \frac{4\pi}{\lambda}\text{Im}(N_{\text{eff}})$. The coupling between the optical field in the port waveguide and in the ring can be described by the field amplitude transmission τ through the coupler and the coupling constant κ. In case of lossless coupling the total energy is conserved and therefore $|\tau|^2 + |\kappa|^2 = 1$. Using these definitions, the transmission of a microring is given by [31, 39]

$$T(\phi) = \left|\frac{E_{\text{out}}}{E_{\text{in}}}\right|^2 = \frac{(\sigma - \tau)^2 + 4\sigma\tau\sin^2(\phi/2)}{(1-\sigma\tau)^2 + 4\sigma\tau\sin^2(\phi/2)}. \tag{1.28}$$

1.3. Electro-optic waveguide modulators

Minima in the transmitted power correspond to constructive interference in the cavity, i.e. to the resonance condition which occurs if

$$\phi = k_0 L n_\text{eff} = 2\pi m, \tag{1.29}$$

where m is an integer. In this case the transmitted light is

$$T(2\pi m) = \frac{(\sigma - \tau)^2}{(1 - \sigma\tau)^2}, \tag{1.30}$$

which is minimum (ideally zero) for the critical coupling condition $\tau = \sigma$. In analogy to a Fabry Perot resonator, the transmission of a microring can be written as [31, 39]

$$T(\phi) = 1 - \frac{T_R}{\left(1 + \frac{4\mathcal{F}^2}{\pi^2} \sin^2(\phi/2)\right)}, \tag{1.31}$$

where $T_R = 1 - T(2\pi m) = (1 - \sigma^2)(1 - \tau^2)/(1 - \sigma\tau)^2$ is the maximal transmission reduction of a microring achievable under resonance conditions and $\mathcal{F} = \pi\sqrt{\sigma\tau}/(1 - \sigma\tau)$ is defined as the finesse $\mathcal{F}$. The finesse relates the free spectral range to the resonance linewidth, which is defined as full width at half maximum $\delta\phi$,

$$\mathcal{F} \approx \frac{2\pi}{\delta\phi}. \tag{1.32}$$

Some important relations for microring resonators are summarized in the following:

- The **free spectral range** $\Delta\lambda_\text{FSR}$, which is an important quantity of a microring resonator, is defined as the spacing between two adjacent resonance wavelengths [31]

$$\Delta\lambda_\text{FSR} \approx \frac{\lambda^2}{L n_\text{eff,g}}, \tag{1.33}$$

where $n_\text{eff,g} = n_\text{eff} - \lambda \frac{\partial n_\text{eff}}{\partial \lambda}$ is the group effective index.

- Expressed in terms of wavelength instead of phase, the **finesse** $\mathcal{F}$ is given by

$$\mathcal{F} = \frac{\Delta\lambda_\text{FSR}}{\delta\lambda_\text{FWHM}}, \tag{1.34}$$

where $\delta\lambda_\text{FWHM}$ is the spectral width of the resonance peak (full width at half maximum).

- The **quality** of a cavity is often described by the **Q** factor of a resonator, which is proportional to the ratio of stored energy in the cavity divided by the loss per roundtrip

$$Q = \frac{\lambda}{\delta\lambda_\text{FWHM}}. \tag{1.35}$$

- The photon lifetime of an unloaded micro ring resonator (absence of a straight waveguide and therefore no coupling losses) is defined as

$$\tau_p = \frac{n_\text{eff,g}}{\alpha c}. \tag{1.36}$$

1. Introduction

In case of weak and critical coupling $\tau = \sigma \approx 1$, some of the quantities introduced can be simplified to [40]

$$\mathcal{F} \approx \frac{\pi}{\alpha L} \quad \text{and} \quad Q \approx \frac{\pi n_{\text{eff,g}}}{\alpha \lambda} = \pi \frac{\tau_p}{T}, \tag{1.37}$$

where $T = 1/\nu$ is the optical field period. Therefore, in a different interpretation the quality factor Q is a measure of the photon cavity lifetime τ_p with respect to the light oscillatory period T. A detailed study of the important topic of bending losses as a function of the ring radius as well as the index contrast between waveguide core and substrate/cladding are given in Ref. [40].

Electro-optic modulation using microring resonators

The tuning of the resonance wavelength of a microring resonance offers several very interesting potential applications, such as switching between different channels for dense wavelength devision multiplexing in modern telecommunication systems or modulation of the transmitted intensity. Since for these applications fast switching is required, electro-optic active rings, particularly those made of organic materials, are of great interest.

Via the electro-optic effect, the effective index of the ring is changed and as a result the optical field experiences an additional phase delay after a round trip, which is in analogy to Eq. (1.26) given by

$$\Delta \phi = k_0 \Delta n_{\text{eff}} L = \frac{\pi}{\lambda} n^3 r \frac{\partial n_{\text{eff}}}{\partial n} \frac{L}{G} V, \tag{1.38}$$

where G is again the distance between the electrodes, assuming that the modulation field is homogenous. Figure 1.4 shows the transmission of a microring as a function of the applied voltage. Indicated is as well a sinusoidal modulation voltage which results in a modulation of the transmitted intensity.

Comparison between Mach-Zehnder and microring resonators

Mach-Zehnder modulators are commonly used for electro-optic modulation and are commercially available with transmission rates of 40 Gbit/s. On the other hand, microrings have attracted a lot of attention during the last years due to their interesting transmission characteristics, nevertheless active tunable microrings are not yet commercially available. A ring and a Mach-Zehnder modulator with the corresponding transmission curves as a function of the applied voltage are shown in Fig. 1.4. In order to compare both types of structures with respect to electro-optic amplitude modulation, the most important modulator properties are listed in Table 1.3 and will be discussed in the following.

- **Maximal slope** and **working point**: Large signal switching by low voltage modulation is possible by setting the working point, where the slope of the transmission curve is the largest. The maximal slope $\partial T/\partial V$ of the transmission curve of a Mach-Zehnder modulator is the largest at 50% transmission. For microring modulators in case of weak and critical coupling a finesse of about $\mathcal{F} = 5$ is required, in order to have the same response in dependence on the applied voltage. Both, the working point as well as

1.4. Advanced modulation formats

the maximal slope are dependent on the finesse of the microring resonator. Microring resonators with finesse above five are therefore advantageous to Mach-Zehnder modulators with respect to their sensitivity to the applied field.

- **Half-wave voltage**: Since for a microring resonator a half-wave voltage in common sense does not exist, an equivalent half-wave voltage $V_\pi^{eq} = \pi/2(|\partial T/\partial V|_{max})^{-1}$ is defined in analogy to the one of a Mach-Zehnder modulator. As well for the half-wave voltage a finesse larger than five is required in order to have a better device performance compared to Mach-Zehnder modulators.

Table 1.3: Main modulator parameters of a microring resonator and a Mach-Zehnder modulator [39].

			Microring	Mach-Zehnder
Transmission	$T(\Delta\phi)$	=	$1 - \dfrac{T_R}{1+\frac{4\mathcal{F}^2}{\pi^2}\sin^2(\Delta\phi/2)}$	$\cos^2(\Delta\phi/2)$
Phase delay	$\Delta\phi$	=	$\frac{\pi}{\lambda}n^3 r \frac{\partial n_{\text{eff}}}{\partial n}\frac{L}{G}V$	$2\frac{\pi}{\lambda}n^3 r \frac{\partial n_{\text{eff}}}{\partial n}\frac{L}{G}V$
Maximal slope	$\left\|\frac{\partial T}{\partial \Delta\phi}\right\|_{max}$	=	$\frac{3\sqrt{3}}{8}T_R\frac{\mathcal{F}}{\pi} \approx \frac{\mathcal{F}}{5}$	$\frac{1}{2}$
Working point	ϕ_p	=	$\frac{1}{\sqrt{3}}\frac{\pi}{\mathcal{F}}$	$\frac{\pi}{2}$
Half-wave voltage	V_π^{eq}	=	$\left\|\frac{\partial T}{\partial \Delta\phi}\right\|_{max}^{-1}\frac{\lambda G}{2n^3 rL}\frac{\partial n}{\partial n_{\text{eff}}}$	$\frac{\lambda G}{2n^3 rL}\frac{\partial n}{\partial n_{\text{eff}}}$

In conclusion, microring modulators potentially require lower driving voltages as compared to Mach-Zehnder modulators. They can also be integrated on much smaller footprints giving a high packing density and they have been shown to have better velocity mismatch tolerances and therefore larger bandwidths than Mach-Zehnder modulators. Traveling-wave microring modulators, which have not been discussed here, show a high modulation efficiency at frequencies around multiples of the free spectral range. So far modulation at 165 GHz has been demonstrated [17], which makes microring resonators highly suited for applications, where an enhanced modulation response is required.

1.4 Advanced modulation formats

1.4.1 High speed optical modulator components

As discussed in section 1.1 the probably most effective kind of modulator is an electro-optic modulator, where a signal voltage changes the refractive index of an electro-optic waveguide, modulating the phase of the guided light. Amplitude modulation can be obtained by generating a phase difference between light in two coherent waveguides and then combining to produce

1. Introduction

constructive or destructive interference as introduced in section 1.3. Such a simple intensity modulation, also called on-off keying (OOK), is the most commonly used modulation format to generate a digital bit stream consisting of a sequence of 0 and 1 bits. However, the achievable spectral efficiency is roughly 1 bit/s per Hz of optical bandwidth for such simple modulation techniques, where the signal power is changing between two levels. Up until a few years ago, optical communication systems primarily employed conventional OOK signals. Especially the emergence of Erbium-doped fiber amplifiers offered enough potential to increase the fiber transmission capacity by using simple intensity modulation systems. Recently, a number of advanced optical modulation formats have attracted considerable attention to increase the fiber capacity toward higher bit rates requested by telecom operators [41, 42]. As an alternative to amplitude modulation formats there is a growing interest in phase modulation formats in order to reach higher spectral efficiencies. Instead of describing the digital information by discrete optical power levels as in case of intensity modulation, the digital signal can also be represented by the phase of an optical carrier, which is commonly referred to as optical phase shift keying (PSK). Since in the detection receiver there is usually no absolute phase reference available, the phase of the preceding bit is used as a relative phase reference. Hence, the digital information is encoded in optical phase changes between consecutive bits and the modulation format is therefore referred to as differential phase shift keying (DPSK). The most simple example for DPSK is differential binary phase shift keying (DBPSK), where two phases are used which are separated by π. Analogously to a two level intensity modulation system DBPSK is able to modulate at 1 bit/symbol. However, there are true multilevel phase modulation formats, which are more efficient than binary signaling. For example, differential quadrature phase shift keying (DQPSK) uses four phases (0, $\pi/2$, π, $3\pi/2$) to encode the signal information. With four phases, DQPSK can encode two bits per symbol, which is twice the rate of DBPSK. In fact, any number of phases may be used to construct a PSK constellation but 8-PSK is usually the highest order deployed. With more than 8 phases, the error-rate becomes too high for practical applications [43].

Table 1.4: Present fiber optic transmission records (records are printed in bold face, see text for explanation).

Cap. (Tbit/s)	Dist. (km)	Cap.·Dist. (Pbit/s·km)	Sp. Eff. (bit/s/Hz)	Nr. of channels	Ch. rate (Gbit/s)	Year	Ref.
32	580	18.56	-	320	114	2009	[44]
1	**9612**	9.6	-	10	111	2009	[45]
13.5	6248	**84.3**	2	135	111	2009	[45]
0.5	240	0.12	**7.0**	8	65.1	2009	[46]
2.56	160	0.41	-	1	**2560**	2006	[47]
5.1	-	-	-	1	**5100**	2009	[48]

1.4. Advanced modulation formats

In addition of adding more levels into basic modulation formats, more complicated formats can be constructed by combining two components, for example, amplitude shift keying and PSK. Since higher values of spectral efficiency can be reached with advanced modulation formats, it is not surprising that many of the recent long-haul transmission records were achieved by systems based on high-order modulation formats. For example, the record per fiber capacity of 32 Tbit/s could be achieved by transmitting 320 wavelength division multiplexed channels with two polarization-multiplexed 114 Gbit/s return-to-zero-8-quadrature amplitude modulation (RZ-8-QAM) signals per channel over 580 km [44]. In a single wavelength channel, 2.56 Tbit/s could be transmitted over 160 km dispersion managed fiber by combining the techniques of optical time division multiplexing, polarization division multiplexing and DQPSK modulation [47], and very recently record single-channel data rates of up to 5.1 Tbit/s by using 16-QAM signals were reported [48], but details such as the transmission distance or the spectral efficiency were not provided. An $8 \cdot 65.1$ Gbit/s polarization- and orthogonal frequency division multiplexed transmission over 240 km standard single mode fiber with the record spectral efficiency of 7.0 bit/s/Hz was demonstrated in Ref. [46], which is the highest spectral efficiency in wavelength division multiplexed transmission with a channel bit rate larger than 40 Gbit/s. A record capacity·distance product of 84.3 Pbit/s·km could be achieved by transmitting 135 channels at 111 Gbit/s over 6248 km [45]. In the same work also the record transmission distance of 9612 km for 100 Gbit/s channels at 1 Tbit/s capacity was reported. These records are summarized in Table 1.4. Since high-order modulation formats are generally seen as a promising way of increasing the capacity of optical fiber transmission systems, we briefly introduce the fundamental concepts of different modulation formats in the following sections.

1.4.2 Non-return-to-zero and return-to-zero on-off keying

In the case of OOK, two very popular choices exist for formatting the bit stream, which are shown in Fig. 1.5a) and b) and known as the non-return-to-zero (NRZ) and return-to-zero (RZ) format respectively. In the RZ format, each bit 1 optical power value is shorter than the bit slot and the amplitude returns to zero before the bit duration is over. In the NRZ format, the optical power value does not drop to zero neither throughout the bit 1 slot nor between two or more successive 1 bits.

To generate a NRZ signal a Mach-Zehnder intensity modulator can be used, which converts an OOK electrical signal into an OOK optical signal. Usually, the Mach-Zender modulator is biased to its quadrature point and is driven from minimum to maximum transmission with the switching voltage V_π as illustrated in Fig. 1.6a.

Often, optical signals with RZ pulse shape are generated by generating a NRZ optical signal as described in the previous paragraph and by subsequently carving RZ pulses with the same data-rate as the electrical signal by cascading another intensity modulator [49, 50]; for example by using a Mach-Zehnder modulator, which is also operated at the quadrature point and driven

1. Introduction

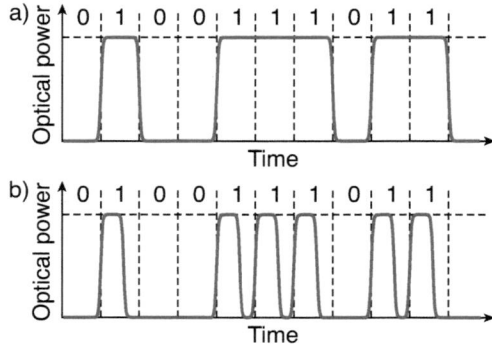

Figure 1.5: Temporal intensity waveform of OOK signals, a) non-return-to-zero and b) return-to-zero.

with a sinusoidal electric driving signal with a peak-to-peak amplitude of V_π and a frequency corresponding to the symbol rate. Since optical pulse distortions (such as chirp) have severe impact on optical fiber transmission performance, the perfection of optical pulse carvers is an often addressed issue (for detailed information on carving techniques see e.g. [49]). RZ optical signals have often been found to be more tolerant to nonlinear transmission degradations than NRZ modulation [49].

1.4.3 IQ-modulator

The basic building block for many phase modulation schemes is the so called IQ-modulator with two nested Mach-Zehnder modulator structures. The optical IQ-modulator is beside the phase modulator and the Mach-Zehnder modulator generally regarded as a third fundamental modulator structure and is also commercially available in an integrated form, for instance from Covega corporation [10]; in the following its operation principle is briefly described. In an optical IQ-modulator the incoming light is equally split into two arms, the in-phase (I) and the quadrature (Q) arm, as illustrated in Fig. 1.7a. In both paths a modulation is performed by operating the respective Mach-Zehnder modulator with a special operation principle. As detailed above for simple intensity modulation, a Mach-Zehnder modulator can be operated at the quadrature point and a peak-to-peak modulation of V_π (see Fig. 1.6a). When the Mach-Zehnder modulator is operated at the minimum transmission point as shown in Fig. 1.6b with a peak-to-peak modulation of $2V_\pi$, a phase shift or π occurs when crossing the minimum transmission point. Now, for the optical IQ-modulator the Mach-Zehnder modulators in the I and Q arm are operated at this minimum transmission point. By adjusting a relative phase shift of $\pi/2$ in one arm, for example with the help of an additional phase modulator, any constellation point can be reached in the complex IQ-plane after recombining the light of both branches (see Fig. 1.7b).

1.4. Advanced modulation formats

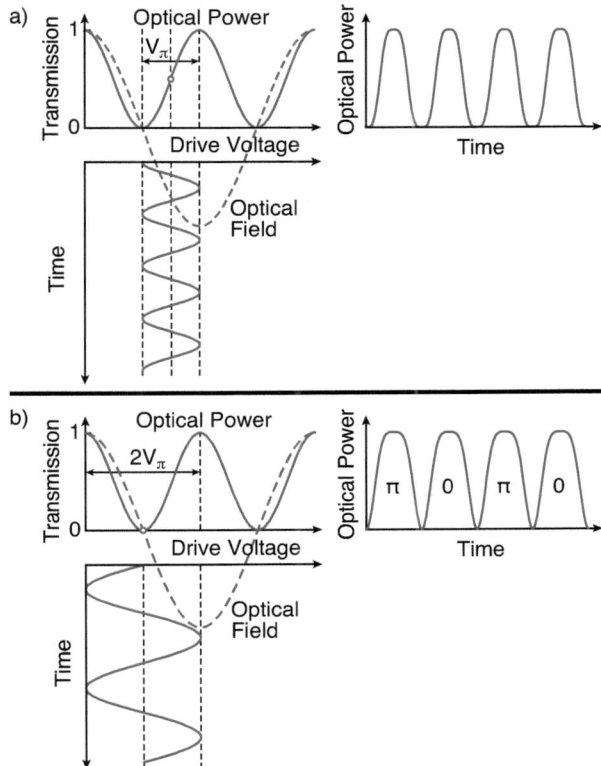

Figure 1.6: Operating a Mach-Zehnder modulator at a) the quadrature point and b) the minimum transmission point. The Mach-Zender modulator bias points are indicated by open circles. Since the phase of the optical field (dashed curve) changes its sign upon transitioning through a minimum in the Mach-Zehnder intensity transmission curve, two neighboring intensity transmission maxima have an optical phase difference of π in b).

In the following the electrical field for the optical output signal is analytically derived in order to better understand the optical IQ-modulator behavior. For the IQ-modulator the induced phase differences of the Mach-Zehnder modulators in the upper and lower path are according to Eq. (1.22)

$$\Delta\phi_I(t) = \frac{u_I(t)}{V_\pi}\pi, \qquad \Delta\phi_Q(t) = \frac{u_Q(t)}{V_\pi}\pi. \tag{1.39}$$

When setting the driving voltage of the phase modulator to $u_{PM} = -V_\pi/2$, the field transfer function of the IQ-modulator can be expressed as [51]

$$\frac{E_{\text{out}}(t)}{E_{\text{in}}(t)} = \frac{1}{2}\cos\left(\frac{\Delta\phi_I(t)}{2}\right) + i\frac{1}{2}\cos\left(\frac{\Delta\phi_Q(t)}{2}\right). \tag{1.40}$$

1. Introduction

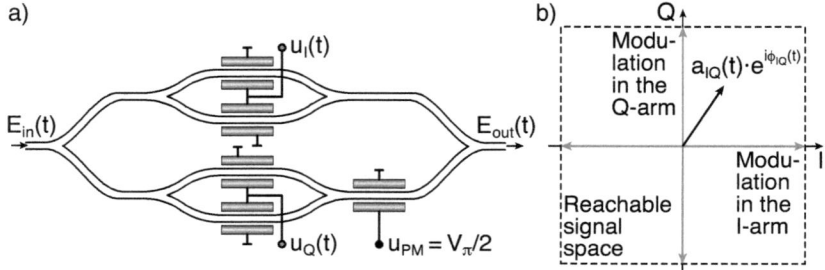

Figure 1.7: The IQ-modulator shown in a) consists of 2 Mach-Zehnder modulators nested in another Mach-Zehnder structure. The phase of the signal passed through the lower modulator, is shifted by $\pi/2$ compared to the one of the upper modulator. With the interference produced at the output of the outer Mach Zehnder structure, any constellation point in the signal space shown in b) can be reached.

By using Eqs. (1.39) and (1.40), the amplitude modulation $a_{IQ}(t)$ and the phase modulation $\phi_{IQ}(t)$, performed by the IQ-modulator, can be calculated as [51]

$$a_{IQ}(t) = \left|\frac{E_{out}(t)}{E_{in}(t)}\right| = \frac{1}{2}\sqrt{\cos^2\left(\frac{u_I(t)}{2V_\pi}\pi\right) + \cos^2\left(\frac{u_Q(t)}{2V_\pi}\pi\right)} \quad (1.41)$$

$$\phi_{IQ}(t) = \tan^{-1}\left(\cos\left(\frac{u_Q(t)}{2V_\pi}\pi\right)\bigg/\cos\left(\frac{u_I(t)}{2V_\pi}\pi\right)\right). \quad (1.42)$$

In Eq. (1.42) the argument of the complex value in the signal space is being calculated from the real and imaginary parts. For representing the various states in the signal space taken by the modulated symbol, a so called constellation diagram is commonly used. For, example, the constellation diagram of the already introduced OOK is depicted in Fig. 1.8a.

1.4.4 DBPSK

According to the reachable signal space for an optical IQ-modulator (see Fig. 1.7b), a single IQ-modulator alone would be sufficient to generate arbitrary DPSK signals. However, the IQ-modulator is not the best choice for the generation of the very simple DBPSK signals, because the two constellation points have the same projection onto the Q-axis (see Fig. 1.8b). Therefore, more effective setups to generate DBPSK signals were developed; the most prominent two configurations are presented below.

One commonly used DBPSK generation setup is depicted in Fig. 1.9a and consists of a phase modulator. The phase modulator only modulates the phase of the optical field while the signal intensity is constant. Since the signal optical power is always constant, this phase modulation format is referred to as NRZ-DBPSK. However, the phase modulation does not occur instantaneously, a phase modulator inevitable introduces chirp across bit transitions [50]. Therefore, again a pulse carver can be used to carve pulses out of the phase-modulated signal, thus generating RZ-DBPSK (see Fig. 1.9a).

1.4. Advanced modulation formats

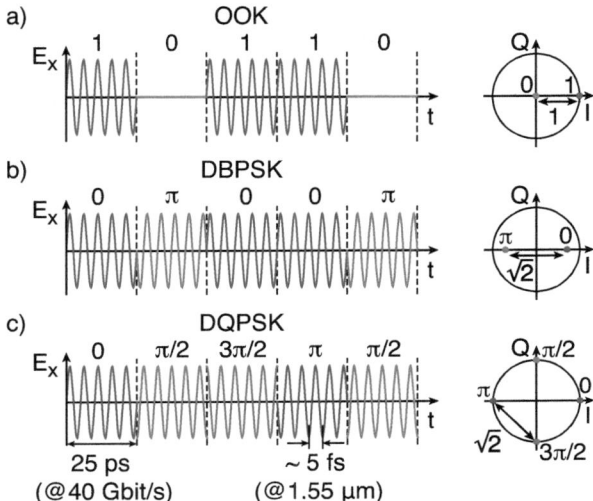

Figure 1.8: The modulated electrical field as a function of time (left) as well as the constellation diagram and symbol distance (right) is represented for three modulation formats: a) OOK, b) DBPSK and c) DQPSK. The oscillation period of the electrical field is around 5 fs for a fiber optic carrier wavelength of about 1.55 µm. Thus, a symbol duration of 25 ps, which corresponds to a signal data rate of 40 Gbit/s, contains approximately 5000 electric field oscillations. Here, only a few oscillations per symbol slot are depicted for clarity. The corresponding constellation diagrams (right) for the different formats are drawn with the same scale and by assuming a constant energy per bit for each modulation format.

A second way of generating optical DBPSK signals is to use a Mach-Zehnder modulator for phase modulation as shown in Fig. 1.9b. The Mach-Zehnder modulator has to be biased at its transmission null and has to be driven at twice the half-wave voltage V_π as discussed earlier and as illustrated in Fig. 1.6b. By transitioning through a minimum in the Mach-Zehnder power transmission curve, the phase of the optical field changes its sign and a phase shift of π is obtained. Nevertheless, the phase modulation with a Mach-Zehnder modulator is accompanied by an amplitude modulation at the transition of two bits. These intensity dips are of reduced importance especially for RZ-DBPSK, where a pulse carver can be used to cut out the amplitude modulation free portions of the bits (see Fig. 1.9b). A detailed comparison between phase modulator and Mach-Zehnder modulator based DBPSK transmitters is given in Ref. [50].

1.4.5 DQPSK with a comparison to OOK and DBPSK

DQPSK is a true multilevel modulation format that has received considerable attention in optical communication in recent years. DQPSK uses four points on the constellation diagram, equispaced around a circle (see Fig. 1.8c). With four phases, DQPSK can encode two bits

33

1. Introduction

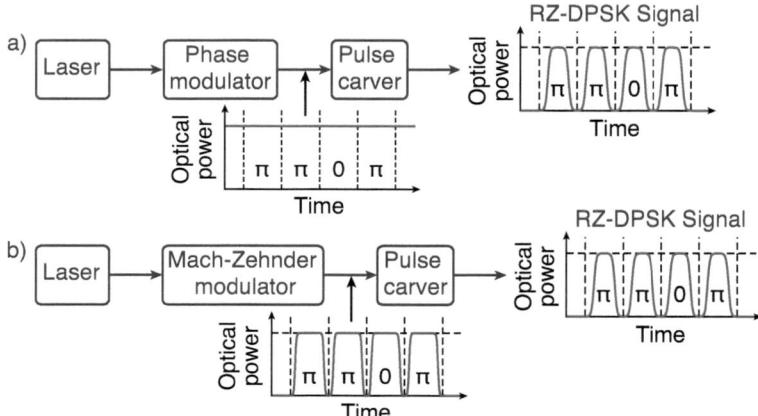

Figure 1.9: Two typical DBPSK transmitters. a) Implementation with a phase modulator. b) Implementation with a Mach-Zehnder modulator.

per symbol, which is twice the rate of DBPSK. A DQPSK transmitter is most conveniently implemented by using the IQ-modulator discussed above. The IQ-modulator transmitter structure requires only binary electronic drive signals, which are much easier to generate at high speeds than for example multilevel electrical drive waveforms.

With the help of constellation diagrams various modulation formats can easily be compared by evaluating the distance between two symbols in the diagram. The constellation diagram for OOK depicted in Fig. 1.8a shows two points plotted along the horizontal line of the diagram, where one point is in the center of the circle and the other on the circle to reflect the low intensity symbol "0" and the high intensity symbol "1" respectively. For DBPSK and DQPSK the signal amplitude is constant but the phase reaches different values, which is illustrated by regularly dispatched states on the trigonometric circle. The constellation diagrams in Fig. 1.8 are drawn in such a way, that an equal average optical power per bit is obtained for each modulation format. For OOK format, one bit can either have an amplitude of 1 or of 0 and the distance between the two possible states is consequently 1. The average power is however $1/2$, since half of the bits have an intensity of 1 while the other bits have an intensity of 0 [52]. For DBPSK the amplitude of each bit is either $\sqrt{2}/2$ or $-\sqrt{2}/2$ to obtain the same average power of $1/2$ as for the OOK signal [52] (note that the intensity of each symbol is $1/2$, because the signal intensity is equal to the square of the signal amplitude). The distance between the two symbols is therefore $\sqrt{2}$ larger than for OOK. Hence, only half the average optical power is needed for DBPSK as compared to OOK to achieve the same symbol distance. On the other hand, this accounts for a 3 dB improved tolerance to optical noise, which is one of the main advantages of DBPSK over OOK [50]. With the DQPSK modulation format, two bits are encoded within each symbol. To achieve an average power per bit of $1/2$, the average power per symbol has to be 1. On the constellation diagram, this translates into symbols located on the circle with radius 1. The distance between each symbol is equal to $\sqrt{2}$, the same as for DBPSK, which indicates that the tolerance to noise for DQPSK can be as good as for DBPSK at twice the capacity [52].

1.4.6 PSK receiver

A simple detector based on a photodiode is usually used to convert the optical signal power into an electrical signal. Photodiodes, however, are insensitive to the phase of the light, therefore an optical pre-processing is necessary in direct-detection receivers to demodulate PSK signals. The optical component, commonly used for this purpose is a delay line interferometer. As illustrated in Fig. 1.10, the light is split by a 3-dB coupler into two arms (i.e. a 50/50 beamsplitter). In one arm, the optical signal is delayed by one symbol duration T (note that DQPSK encodes two bits per symbol) and additionally a phase shift ϕ_{DLI} can be accomplished in one of the branches before the light in the two arms is recombined in a second 3-dB coupler. To describe a single lossless 3-dB coupler, the general coupled first-order mode equations [26] can be used, which dramatically simplify and can be expressed in matrix notation as

$$\begin{pmatrix} \tilde{E}_{\text{out}_1} \\ \tilde{E}_{\text{out}_2} \end{pmatrix} = \frac{1}{\sqrt{2}} \begin{pmatrix} 1 & i \\ i & 1 \end{pmatrix} \begin{pmatrix} E_{\text{in}_1} \\ E_{\text{in}_2} \end{pmatrix}, \tag{1.43}$$

where for the various electric fields the nomenclature introduced in Fig. 1.10 is used. When only the signal E_{in_1} at the upper port is present at the inputs of the delay line interferometer, the two fields obtained at the output ports are given by

$$\begin{pmatrix} E_{\text{out}_1} \\ E_{\text{out}_2} \end{pmatrix} = \frac{1}{2} \begin{pmatrix} 1 & i \\ i & 1 \end{pmatrix} \begin{pmatrix} e^{-i\omega_0 T} & 0 \\ 0 & e^{i\phi_{DLI}} \end{pmatrix} \begin{pmatrix} 1 & i \\ i & 1 \end{pmatrix} \begin{pmatrix} E_{\text{in}_1} \\ 0 \end{pmatrix} \tag{1.44}$$

$$= \begin{pmatrix} \frac{1}{2}E_{\text{in}_1}(t-T) - \frac{1}{2}E_{\text{in}_1}(t)e^{i\phi_{DLI}} \\ i\frac{1}{2}E_{\text{in}_1}(t-T) + i\frac{1}{2}E_{\text{in}_1}(t)e^{i\phi_{DLI}} \end{pmatrix}, \tag{1.45}$$

where ω_0 is the frequency of the electric field. For a delay line input signal

$$E_{\text{in}_1}(t) = \sqrt{P_0} \cdot e^{i(\omega_0 t + \phi_0)} \cdot e^{i\phi(t)}, \tag{1.46}$$

where $\sqrt{P_0}$ is the field amplitude and $\phi(t)$ is the modulation phase, the optical power at the delay line interferometer outputs is given by

$$P_{\text{out}_1}(t) = E_{\text{out}_1}(t) \cdot E^*_{\text{out}_1}(t) = \frac{1}{2}(P_0 - P_0 \cos(\Delta\phi(t) + \phi_{DLI})) \tag{1.47}$$

$$P_{\text{out}_2}(t) = E_{\text{out}_2}(t) \cdot E^*_{\text{out}_2}(t) = \frac{1}{2}(P_0 + P_0 \cos(\Delta\phi(t) + \phi_{DLI})), \tag{1.48}$$

where the modulation phase difference $\Delta\phi(t)$ between two consecutive symbols is

$$\Delta\phi(t) = \phi(t) - \phi(t-T). \tag{1.49}$$

1. Introduction

In the derivation of Eqs. (1.47) and (1.48) it has been assumed that an integer number of wavelength fits into the symbol time ($\omega_0 T = 2\pi m$, $m \in \mathbb{N}_0$) and ϕ_0 takes into account an extra relative phase shift. The delay line interferometer output signals can be used for the evaluation of $\Delta\phi(t)$, when an appropriate interferometer phase shift ϕ_{DLI} is adjusted. For example, to demodulate a DBPSK-encoded bit stream no additional interferometer phase shift ϕ_{DLI} is required beside the 1 bit delay between the two arms of the interferometer. If two adjacent bits have a phase difference $\Delta\phi(t) = 0$ ($\Delta\phi(t) = \pi$), this leads to the absence (presence) of optical signal power at the upper delay line interferometer exit and vice versa for the lower interferometer exit. Therefore, in principal one of the delay line interferometer output ports is sufficient to demodulate DBPSK signals, however usually a balanced detector is employed to increase the signal to noise ratio.

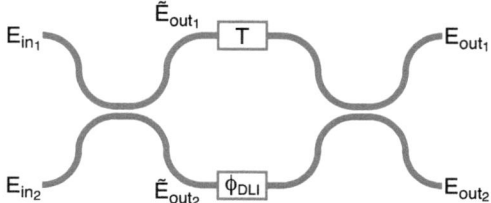

Figure 1.10: In a delay line interferometer, an incoming optical signal is first split into two equal-intensity beams with the help of a 3-dB optical coupler. Subsequently, one of the beams in the Mach Zehnder interferometer is delayed by an optical path difference corresponding to a time delay T of one symbol and the other beam experiences an additional phase shift ϕ_{DLI} before the light of both arms is recombined in a second 3-dB coupler.

For DQPSK, the signal phase can reach four values, 0, $\pi/2$, π, $3\pi/2$. To demodulate such DQPSK signals two asymmetric delay line interferometers are required in a direct-detection interferometric receiver (see Fig. 1.11). The output signals of the receiver in Fig. 1.11 can be calculated similarly as for a single delay line interferometer. Under the assumption that only the input signal E_{in_1} at the upper branch of the receiver is present, the output signals are given by

$$E_{\text{out}_1} = \frac{1}{2\sqrt{2}}(E_{\text{in}_1}(t-T) - E_{\text{in}_1}(t)e^{i\phi_{\text{DLI}_1}}) \tag{1.50}$$

$$E_{\text{out}_2} = \frac{1}{2\sqrt{2}}(iE_{\text{in}_1}(t-T) + iE_{\text{in}_1}(t)e^{i\phi_{\text{DLI}_1}}) \tag{1.51}$$

$$E_{\text{out}_3} = \frac{1}{2\sqrt{2}}(-E_{\text{in}_1}(t-T) - E_{\text{in}_1}(t)e^{i\phi_{\text{DLI}_2}}) \tag{1.52}$$

$$E_{\text{out}_4} = \frac{1}{2\sqrt{2}}(-iE_{\text{in}_1}(t-T) + iE_{\text{in}_1}(t)e^{i\phi_{\text{DLI}_2}}), \tag{1.53}$$

where the output signals of the upper interferometer are E_{out_1} and E_{out_2} for upper and lower branches respectively and the output signal of the lower interferometer are E_{out_3} and E_{out_4} for

1.4. Advanced modulation formats

upper and lower branches respectively; ϕ_{DLI_1} and ϕ_{DLI_2} are the phase shifts of the upper and lower delay line interferometer respectively. For an input signal E_{in_1} given as in Eq. (1.46), the corresponding output powers at the four outputs are given by

$$P_{\mathrm{out}_1} = \frac{1}{4}P_0(1 - \cos(\Delta\phi(t) + \phi_{\mathrm{DLI}_1})) \qquad (1.54)$$

$$P_{\mathrm{out}_2} = \frac{1}{4}P_0(1 + \cos(\Delta\phi(t) + \phi_{\mathrm{DLI}_1})) \qquad (1.55)$$

$$P_{\mathrm{out}_3} = \frac{1}{4}P_0(1 + \cos(\Delta\phi(t) + \phi_{\mathrm{DLI}_2})) \qquad (1.56)$$

$$P_{\mathrm{out}_4} = \frac{1}{4}P_0(1 - \cos(\Delta\phi(t) + \phi_{\mathrm{DLI}_2})). \qquad (1.57)$$

By detecting E_{out_1} and E_{out_2} with an upper balanced detector and E_{out_3} and E_{out_4} with a lower balanced detector, we get the photocurrents

$$\tilde{I} = P_{\mathrm{out}_1} - P_{\mathrm{out}_2} \qquad (1.58)$$

$$\tilde{Q} = P_{\mathrm{out}_3} - P_{\mathrm{out}_4}, \qquad (1.59)$$

where in practice the subtraction operations can be performed electronically after photodetection. By adjusting the differential optical phase between the delay line interferometer arms to $\phi_{\mathrm{DLI}_1} = \pi/4$ and $\phi_{\mathrm{DLI}_2} = -\pi/4$ and by using balanced optical detection with 0 as the threshold, the binary in-phase (I) and quadrature (Q) components of the optical DQPSK signal can be detected (i.e. $I = 0$ for $\tilde{I} < 0$, $I = 1$ for $\tilde{I} > 0$ and analogously for Q and $\tilde{Q}$). In Table 1.5 the relationship of the phase of the input optical signal and the balanced binary detection signals I and Q is shown. Thus, output voltage signals are obtainable which correspond to the encoded bits of the data. Both binary as well as quadrature-phase demodulators are commercially available for example from ITF Labs [53] or Optoplex [54].

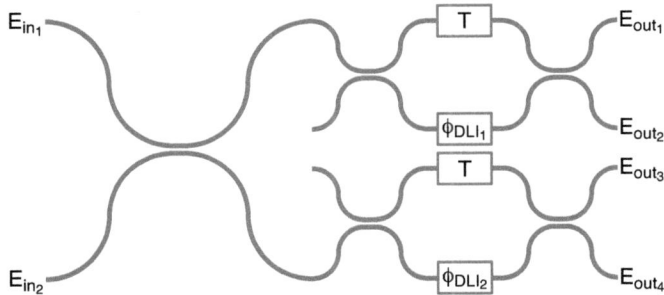

Figure 1.11: A conventional DQPSK receiver consists of two delay line interferometers. Each interferometer has a delay of one symbol period T, but the corresponding additional phase shift ϕ_{DLI_1} and ϕ_{DLI_2} must be chosen differently to decode the in-phase and quadrature component correctly (see text).

To demodulate DQPSK signals also other direct-detection receiver schemes have been proposed, see Ref. [55] for different realization options. Alternatively to direct-detection,

1. Introduction

Table 1.5: Relationship of the relative phase difference of successive symbols and the balanced binary detection signals I and Q, where 0 is used as the threshold value for balanced optical detection.

DQPSK phase transition	I	Q
0	0	1
$\pi/2$	1	1
π	1	0
$3\pi/2$	0	0

coherent detection can be employed. In the case of coherent detection, the phase reference is provided by a local laser within the receiver, which oscillates at the same frequency as the signal frequency. Detailed information about coherent detection can be found for example in Ref. [52].

Note, that very recently techniques of using microring resonators to design DBPSK and DQPSK modulators and demodulators have been reported [56, 57]. The proposed systems offer a smaller chip area and easier fabrication into arrays compared to conventional Mach-Zehnder structures. In the following one of the proposed configurations of using microring resonators to generate DQPSK signals is briefly discussed. For properly designed microring resonators, the resonance peak can be shifted, such that a continuous wave laser source experiences a phase shift of π across a dip in the transmission curve, while having the same power in both phase-states [58]. Therefore a single microring modulator can be used to modulate DBPSK signals (see constellation diagrams in Fig. 1.12). To obtain DQPSK signals, consequently two microring modulators generating independently DBPSK signals, can be incorporated in a single interferometric structure, where an additional microring acts as a phase shifter to induce a phase shift of $\pi/2$ in one of the interferometer arms as shown in Fig. 1.12. DQPSK data formats can be generated this way.

1.5 Summary

As the demand for communication bandwidth in multi-core microprocessors rises, it is expected that metal interconnects will run into problems from excessive power consumption and latency in the foreseeable future and that optical interconnects offer better performance. Replacing electrical interconnections by optical systems, at different levels ranging from rack-to-rack down to chip-to-chip and intra-chip devices, may be the only solution to overcome the problem of bandwidth limiting electrical interconnects in state-of-the-art computer processors. In this context, the development of solutions to efficiently perform switching and modulation of optical signals on a reduced size is of paramount importance.

One possibility to achieve the required level of chip-scale optical integration, is to exploit the particular geometrical-dependent spectral characteristics of resonant optical microcavities.

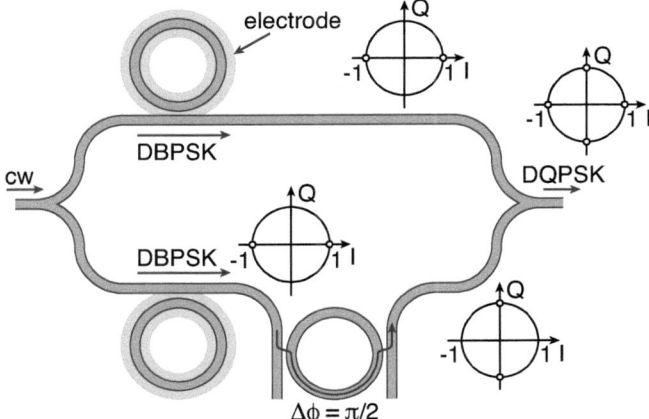

Figure 1.12: DQPSK data can be generated by an interferometric structure, in which two microrings are used as DBPSK modulators, while an additional microring induces a phase shift of $\pi/2$ between the two arms of the Mach-Zehnder-like interferometer. The open circles in the constellation diagrams indicate the states taken in the signal space by the modulated data, drawn for different stages in the structure.

Therefore, optical modulators based on ring-like optical cavities are among the most promising components in optoelectronic integration. To achieve modulation speeds in excess of 150 GHz, organic electro-optic materials with their inherently ultra-fast Pockels effect and their almost purely electronic hyperpolarisabilities are highly interesting for electro-optic modulators. In addition, organic materials generally feature a low dispersion of the dielectric constant. Therefore, velocity matching between the optical wave and the modulation field for example in Mach–Zehnder modulators can easier be achieved compared to standard inorganic materials. To further increase the available optical bandwidth in future transmission systems, advanced modulation formats, which can provide enhanced spectral efficiency, have attracted considerable attention in recent years. Usually combinations of optical Mach–Zehnder modulators, microring resonators and phase modulators are used to generate such advanced modulation formats.

1.6 Outline of the thesis

Organic non-centrosymmetric materials are not only of great interest for electro-optic light modulation but also for parametric light generation, frequency conversion or terahertz wave generation. The general requirement for the usage in such nonlinear optical applications is a high second-order nonlinear optical susceptibility of the organic material. Presently the most well developed organic nonlinear optical crystal, which has also been used in several prototype photonic devices, is the stilbazolium salt DAST that exhibits a relatively large nonlinear optical susceptibility $\chi^{(2)} = 420 \pm 110$ pm/V at $\lambda = 1.907$ µm. Despite various investigations and

1. Introduction

studies, the growth of large size high optical quality DAST crystals is still challenging and time consuming. Furthermore, the theoretical limits of the second-order nonlinear optical activity of crystalline materials are still far from being reached. The development of new materials, which are easier to grow and which exhibit larger nonlinear optical coefficients is therefore desired. Various organic compounds have been developed with the aim of optimizing the properties of the individual molecules, as well as improving their arrangement in non-centrosymmetric crystals in order to enhance their nonlinearity. The organic ionic-type crystal DAPSH (trans-4'-(dimethylamino)-N-phenyl-4-stilbazolium hexafluorophosphate) with a stronger electron acceptor than that in DAST has been recently designed. In various experimental and theoretical investigations of the chromophores a considerable increase of the first hyperpolarizability β for the DAPSH chromophore compared to the DAST chromophore was found. In chapter 2 the linear and nonlinear optical properties of DAPSH in bulk single crystals are reported. We demonstrate that DAPSH has a superior non-resonant second-order susceptibility $\chi^{(2)} = 580 \pm 80$ pm/V as compared with all previously reported crystalline materials.

The organic crystal DAPSH has to be grown from solution. On the other hand, melt growth, which is advantageous because of much faster growth rates, cannot be applied since the decomposition temperature of the DAPSH chromophore is below its melting temperature. Therefore, new polyene chromophores, which were configurationally locked to increase their thermal stability, were developed in our laboratory. Among the investigated molecules, DAT2 (2-(3-(2-(4-dimethylaminophenyl)vinyl)-5,5-dimethylcyclohex-2-enylidene)malononitrile) has been found to exhibit a non-centrosymmetric crystal structure and has shown a high tendency to form thin films. Even though the crystalline nonlinearity of DAT2 is not as high as the one of the stilbazolium salts DAST or DAPSH, the possibility to apply melt growth techniques as well as the easier thin film processing makes DAT2 a very promising and attractive material for applications in integrated electro-optics. Indeed, in chapter 3 we report on the fabrication of high-quality single-crystalline high-index-contrast waveguiding structures and demonstration of the electro-optic modulation at the telecommunication wavelength 1.55 μm in DAT2 waveguides. The crystals were grown from the melt inside straight waveguide channels and were efficiently protected against chemical and mechanical damage. In addition the developed melt-based channel growth technique is also very suitable for the growth of sub 30 nm thick crystalline nanowires and nanosheets.

Based on the melt based channel growth technique, integrated electro-optic Mach-Zehnder modulators and microring resonators were fabricated and are reported in chapter 4. The appropriately shaped waveguide channels for subsequent crystallization of the melt of the organic material therein were obtained by standard optical lithographic techniques and wafer bonding. The geometrical waveguide dimensions, which are especially relevant in microring resonators, could be accurately controlled, since photolithographic processing was only applied to the inorganic substrate and cover wafers but not directly to the organic thin film crystals. In order to reduce the problem of thermal back-dilatation inherent to melt growth methods

1.6. Outline of the thesis

(i.e. to minimize the crack formation during the cooling of the film down to room tempurature), COANP (2-cyclo-octylamino-5-nitropyridine) with its relatively low melting point was chosen as electro-optic material for the microring fabrication, while we have used DAT2 for the Mach-Zehnder modulators. The half-wave voltage × length product determined in the DAT2 based phase and Mach-Zehnder modulators was found to be in good agreement; we reached 78 ± 2 Vcm for TE-modes and 60 ± 1 Vcm for TM-modes at a wavelength of 1.55 μm. By electro-optically modulating the COANP microring resonators an approximate tunability of 0.13 GHz/V (1.1 pm/V) was found, which is in good agreement with theoretical prediction and comparable to what has been reported for ion-sliced $LiNbO_3$ microring resonators.

In summary, this thesis demonstrates the realization of electro-optic single-crystalline organic phase modulators, Mach-Zehnder modulators as well as microring resonators. In contrast to poled electro-optic polymer waveguide structures, the single-crystalline devices exhibit a superior photochemical and thermal stability. The combination of the beneficial electro-optic characteristics of organic single crystals with the possibility of device size reduction using microring resonators is an important step towards ultra-fast optical data processing in highly-integrated photonic devices.

1. Introduction

CHAPTER 2

Extremely large non-resonant second-order nonlinear optical response in crystals of the stilbazolium salt DAPSH[*]

We report on the extremely large non-resonant quadratic optical nonlinearity of the stilbazolium salt trans-4'-(dimethylamino)-N-phenyl-4-stilbazolium hexafluorophosphate (DAPSH). The phenyl-pyridinium chromophores in DAPSH crystals grown from acetone solution pack with a highly aligned polar order, resulting in a very large birefringence, $\Delta n = 1.17 \pm 0.06$ at $\lambda = 0.83$ µm and $\Delta n = 0.83 \pm 0.04$ at $\lambda = 1.55$ µm. More importantly, this leads to an extremely large diagonal quadratic susceptibility with the nonlinear optical coefficient for second harmonic generation reaching up to $d_{111} = 290 \pm 40$ pm/V at 1.907 µm fundamental wavelength, which presents a considerable improvement with respect to the presently best material DAST (4-N, N-dimethylamino-4'-N'-methyl-stilbazolium tosylate) with $d_{111} = 210 \pm 55$ pm/V at $\lambda = 1.907$ µm. The result is in agreement with the preferential packing of the chromophores and the previous studies demonstrating higher microscopic nonlinearity of the chromophores in DAPSH compared to that of DAST.

2.1 Introduction

In the field of organic nonlinear and electro optics, much effort is directed at the design of new molecules for polymers or crystals with high nonlinearities. Owing to an almost pure electronic origin of their nonlinearities, organic materials are extremely promising for future ultra-fast photonic devices [16, 19, 21, 28, 59–63]. Compared to the widely investigated poled polymers, organic single crystals are advantageous because of superior long-term thermal and photochemical stability combined with a higher chromophore concentration [28, 61–63]. Presently, the most well known and investigated organic nonlinear optical crystal is the stilbazolium salt DAST (4-N, N-dimethylamino-4'-N'-methyl-stilbazolium tosylate) [64, 65]. It

---

[*]This chapter, together with Appendix A has been published in the Journal of the Optical Society of America B: Optical Physics **25**(11), 1786–1793 (2008).

exhibits electro-optical and nonlinear optical figures of merit that are comparable or even higher than those of presently known standard inorganic materials and poled polymers, i.e. nonlinear optical coefficient $d_{111} = 210 \pm 55$ pm/V at $\lambda = 1.907$ µm [66] and electro-optic figure of merit $r_{111}n_1^3 = 530 \pm 60$ pm/V at $\lambda = 1.313$ µm [67]. DAST has been used as a very efficient THz emitter [68, 69] and in integrated electro-optic structures [70–73].

The development of novel nonlinear optical crystals is motivated by the fact that the theoretical limits of the macroscopic second-order nonlinearities in organic crystals are still far from being achieved [28, 62]. During the optimization of the molecular structure a variety of organic compounds have been developed and investigated. Several DAST derivatives with modified counter anions have shown a big potential to increase the nonlinearity through the optimization of the crystal packing and also to improve the crystal growth characteristics [74–77], leading to faster and easier growth of bulk crystals and single crystalline thin films of DAST analogues [78, 79]. On the other hand, N-aryl-pyridinium chromophores have increased first hyperpolarizability values β when compared with their N-methyl counterparts as used in DAST [80, 81]. Among these, the organic ionic-type crystal DAPSH (trans-4'-(dimethylamino)-N-phenyl-4-stilbazolium hexafluorophosphate) was found to be very interesting for nonlinear optical applications [82, 83]. Based upon Stark spectroscopic measurements as well as density functional theory (DFT) and coupled perturbed Hartree-Fock (CPHF) theoretical studies, the non-resonant beta value of the chromophore in DAPSH is about 20-50% larger than that of its counterpart in DAST [82–84]. Also the crystal packing of DAPSH chromophores is expected to be better suited for electro-optic applications due to a smaller angle between the molecular dipoles compared to DAST. Therefore, very high macroscopic nonlinear optical figures of merit are expected and are investigated in this work.

2.2 Crystal structure and sample preparation

DAPSH is an ionic crystal with a positively charged nonlinear optical chromophore N-phenyl-stilbazolium and a hexafluorophosphate anion as shown in Fig. 2.1a). Depending on the growth conditions, DAPSH exhibits several polymorphic forms [85]. Here we consider only the DAPSH-a form that has a noncentrosymmetric structure required for second-order nonlinear optics [85]. DAPSH-a crystals belong to the monoclinic space group Cc (point group m). The molecular cell contains four cation-anion pairs. The lattice parameters are $a = 19.384$ Å, $b = 10.636$ Å and $c = 11.784$ Å with $\alpha = \gamma = 90°$ and $\beta = 125.93°$ [82]. In the point group symmetry m, the dielectric x_2 axis is set parallel to the crystallographic b axis. The orientation of the molecules with respect to the unit cell in the ac plane (mirror plane) is shown in Fig. 2.1b). In this plane, the main charge-transfer axis of the nonlinear optical chromophores makes an angle $\psi = 55.8°$ with the normal to the bc plane. The crystal used in our experiments (see Fig. 2.2a) was grown from acetone solution as described in Ref. [85]. The largest natural face (1,0,0) was polished to $\lambda/4$ surface quality. The Cartesian system illustrated in Fig. 2.1b), 2.1c) and 2.2b) with the

2.2. Crystal structure and sample preparation

x_1 axis parallel to the charge-transfer axis, the x_2 axis parallel to the crystallographic b axis and x_3 perpendicular to x_1 and x_2 coincides with the dielectric system, as confirmed by the refractive index measurements. The thickness of the crystal in Fig. 2.2a) perpendicular to the bc plane was 816 ± 3 µm.

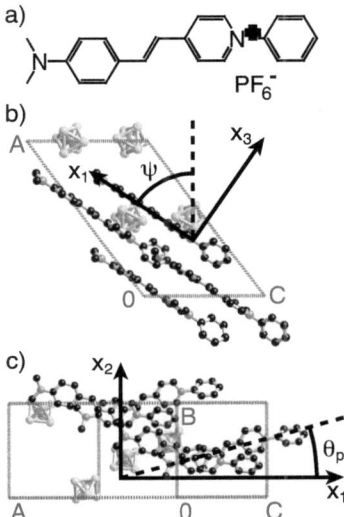

Figure 2.1: a) Molecular diagram of DAPSH with the nonlinear optical active cation (phenyl-stilbazolium) and the negatively charged anion (PF_6). b) A view of the crystal packing along the crystallographic b axis showing the high alignment of the chromophores in the ac plane; ψ is the angle between the polar axis (dielectric x_1 axis) and the normal to the bc plane (dashed line). c) A view along the dielectric axis x_3. An angle $\theta_p = 15.5°$ between the long axis of the molecules and the polar axis is indicated.

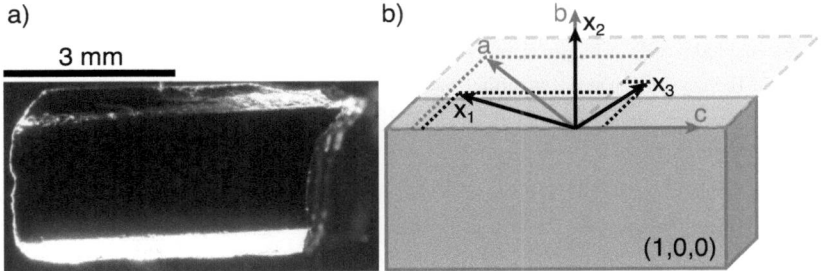

Figure 2.2: a) Dark field microscope image of the crystal used for our measurements, the largest surface shown is (1,0,0). b) Schematic illustration of the orientation of the crystal shown in Fig. 2.2a).

2. Linear and nonlinear optical properties of DAPSH

2.3 Refractive indices

The refractive indices of DAPSH were measured with an interferometric method [28, 86], which is based on changing one optical path length in the Michelson interferometer by rotating the samples. For details of the measurement see Appendix A.2. The accuracy of the results is limited by the error in the orientation of the crystal, by the error in the parallelism of the surfaces of the sample, and by the absolute value of the crystal thickness.

The measured refractive indices of DAPSH are shown in Fig. 2.3. We describe the dispersion of the refractive indices $n(\lambda)$ with a simple Sellmeier single-oscillator model given by

$$n^2(\lambda) = n_0^2 + \frac{q\lambda_0^2}{\lambda^2 - \lambda_0^2} = n_0^2 + \frac{qE^2}{E_0^2 - E^2} \; , \tag{2.1}$$

where $\nu_0 = c/\lambda_0$ is the resonance frequency of the main oscillator, q is the oscillator strength, n_0 is a constant depending on the contributions from all other oscillators. The expression on the right-hand side is the corresponding energy description, with the oscillator energy $E_0 = h\nu_0$. The full lines in Fig. 2.3 are according to Eq. (2.1) obtained with the parameters given in Table 2.1. As also observed for other organic nonlinear optical crystals with highly aligned chromophores, a single-oscillator model gives different resonance wavelengths for the main refractive indices. Because of the weak dispersion in the investigated wavelength range of the refractive index n_3, the corresponding data points were approximated by a constant, i.e. considering $q = 0$. This is in accordance with the direction of the x_3 axis (Fig. 2.1b), which is perpendicular to the charge-transfer axes of the chromophores. In the absorption spectrum of DAPSH powder, two distinct absorption peaks are visible with peak wavelength at 484 ± 15 nm and 607 ± 10 nm. Since the charge-transfer axis of the chromophores is highly aligned along the x_1 axis (Fig. 2.1b), the peak observed at the longer wavelength is considered as the resonance wavelength λ_0 of the corresponding refractive index n_1. The peak at the shorter wavelength corresponds well with the measured dispersion of n_2, but its choice as λ_0 is less justified and therefore considered with a bigger error (see Table 2.1). Two resonance peaks are also observed for DAST in the crystalline state, at about 500 nm and 535 nm. For comparison, the absorption maximum in acetonitrile solution is at $\lambda_{\text{max}} = 470$ nm for the PF$_6$-analogue of DAST and at 504 nm for DAPSH [82].

The results show that DAPSH is highly anisotropic with a birefringence of up to $\Delta n = n_1 - n_3 = 1.17 \pm 0.06$ at $\lambda = 0.83$ µm and $\Delta n = 0.83 \pm 0.04$ at $\lambda = 1.55$ µm in the ac crystallographic plane. This is more than 50% larger than in DAST, for which $\Delta n \simeq 0.74$ at $\lambda = 0.83$ µm and $\Delta n \simeq 0.56$ at $\lambda = 1.55$ µm [67]. The high birefringence Δn can be well understood by the almost parallel alignment of the chromophores and the large electron polarizability along the x_1 axis (long axis of the chromophores) compared to the one perpendicular to the molecular plane (x_3) (Fig. 2.1b).

2.4. Absorption coefficients

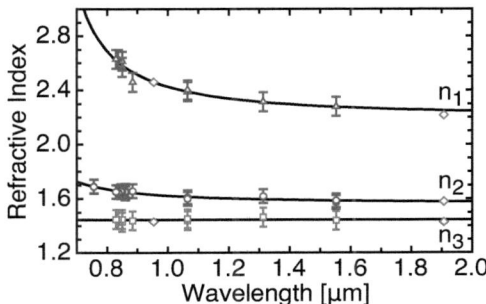

Figure 2.3: Measured refractive index n_1 (open triangles), n_2 (open circles), and n_3 (open squares) of DAPSH as a function of the wavelength λ. The solid curves are Sellmeier oscillator fits using Eq. (2.1). The refractive indices, which were used for the analysis of the Maker fringe data are depicted as open diamonds (see text).

Table 2.1: Sellmeier parameters for the description of the dispersion of the refractive indices (Eq. (2.1)). The given values are the parameters for calculating the refractive indices, while the error indicated is the error range of the measurement. [1] $q = 0$ was assumed for n_3 since the measured data did not allow for a reasonable dispersion fit. [2] the resonance wavelengths were obtained from an absorption spectra measured in transmission using DAPSH powder.

	n_1	n_2	n_3
q	1.728 ± 0.31	0.581 ± 0.23	0^1
n_0	2.208 ± 0.06	1.567 ± 0.03	1.447 ± 0.03
λ_0 [nm]	$607^2 \pm 10$	$484^2 \pm 100$	-
E_0 [eV]	2.04 ± 0.03	2.6 ± 0.5	-

2.4 Absorption coefficients

The transmission of DAPSH was measured for normal incidence for light polarized parallel and perpendicular to the dielectric axis x_2 in the wavelength range from 0.6 µm up to 2 µm using a Perkin Elmer Lambda 9 spectrometer. The absorption coefficients were calculated from the measured transmission curves taking into account Fresnel losses due to multiple reflections at the crystal surfaces. The results are shown in Fig. 2.4. In the near infrared wavelength range absorption bands at about 1.1 µm, 1.4 µm and 1.7 µm can be observed, which correspond to overtones of the C-H stretching vibration, as found in other organic crystals. Small imperfections in the investigated crystal and imperfect polishing might be an explanation for the fact, that the absorption coefficient is larger than 0.7 cm^{-1} for both polarizations over the measured wavelength range. The absorption edge, defined for an absorption of 10 cm^{-1}, is at about 725 nm for both polarizations. In the telecommunication wavelength range around 1.55 µm the measured absorption constant is about $\alpha = 1$ cm^{-1}.

2. Linear and nonlinear optical properties of DAPSH

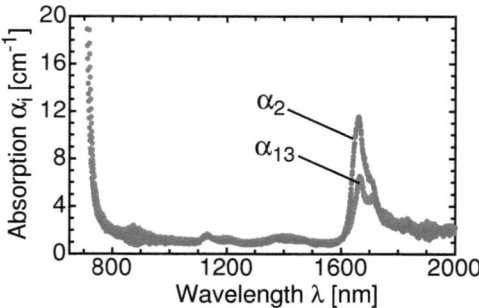

Figure 2.4: Absorption coefficient for eigenpolarizations α_2 and α_{13} as a function of the wavelength λ.

2.5 Nonlinear optical properties

2.5.1 Nonlinear optical tensor for second harmonic generation

The nonlinear optical tensor elements d_{ijk} can be related to the molecular hyperpolarizabilities by the model proposed by Bergman and Crane [87] and applied to the case of organic crystals by Zyss and Oudar [88]. The model neglects contributions of intermolecular interactions to the nonlinearity, except of the local field corrections. This leads to a simple structural dependence of the nonlinear optical susceptibilities for second harmonic generation (SHG)

$$d_{ijk} = \frac{1}{2} N \frac{1}{n(g)} f_i^{2\omega} f_j^{\omega} f_k^{\omega} \cdot \sum_s^{n(g)} \sum_{ijk}^{3} \cos\theta_{il}^s \cos\theta_{jm}^s \cos\theta_{kn}^s \beta_{lmn} , \qquad (2.2)$$

where N is the number of molecules per unit volume, $n(g)$ is the number of equivalent positions in the unit cell, s denotes a site in the unit cell, $f_i^{\omega,2\omega}$ are the local field corrections, β_{ijk} is the first order molecular hyperpolarizability tensor element and $\cos(\theta_{il}^s)$ is the projection of the charge-transfer axis l to the dielectric axis i of the crystal. For dipolar molecules with strong nonlinearities along a single charge-transfer axis, the hyperpolarizability β along this axis is dominant and other contributions can be neglected. Our measurements of the refractive indices showed that the dielectric axis x_1 is oriented along the charge-transfer axis in the ac plane within an error of $\pm 2°$ (see Appendix A.2). Therefore, the angle between the dielectric axis x_3 and the charge-transfer axis in the ac plane is almost $90°$. This results in a projection factor on the x_3 axis which is almost zero and hence the nonlinear optical tensor elements, with one or more indices 3, will be small. Thus the nonlinear optical tensor d_{ijk} in the chosen Cartesian dielectric system is approximately given by

$$d_{ijk} \approx \frac{1}{2} N f_i^{2\omega} f_j^{\omega} f_k^{\omega} \cdot \begin{pmatrix} \beta\cos^3\theta_p & \beta\cos\theta_p\sin^2\theta_p & \ll\beta & 0 & \ll\beta & 0 \\ 0 & 0 & 0 & \ll\beta & 0 & \beta\cos\theta_p\sin^2\theta_p \\ \ll\beta & \ll\beta & \ll\beta & 0 & \ll\beta & 0 \end{pmatrix} . \qquad (2.3)$$

where the matrix elements are given in the usual contracted notation $d_{ijk} = d_{im}$ ($m = 1,\ldots,6$).

2.5.2 Maker fringe measurements

The nonlinear optical tensor elements for second harmonic generation were measured by the standard Maker fringe technique adapted for anisotropic absorbing materials [89], as explained in Appendix A.1.

The tensor elements d_{111} and d_{122} were determined by rotating the sample around the dielectric x_2 axis. In a first measurement d_{122} was determined by type I phase matching, where the fundamental beam was s-polarized and the p-polarized component of the generated second harmonic light was detected (see Fig. 2.5). From the SHG intensity we found $d_{\text{eff}}^{(1)} = 3.05 \pm 0.3$ pm/V, which corresponds to a $d_{122} = 15 \pm 2$ pm/V, assuming $d_{322} = 0$. The measured SHG intensity as a function of the rotation angle depicted in Fig. 2.5 shows a peak width which is somewhat larger than that of the theoretical curve. This effect can be explained by the focusing and/or scattering of the laser beam on the crystal. Therefore the wave vectors **k** have a small distribution in space and are not perfectly unidirectional as assumed for the theoretical curve. In the accessible range of the angle of rotation no other phase matched SHG signal peak was visible outside the rotation range shown in Fig. 2.5, but only the ordinary Maker fringes originating from the angular dependence of the phase mismatch between the fundamental and second harmonic waves. The detailed analysis of these Maker fringes, observed as side ripples in Fig. 2.5, did not bring more information to the nonlinear optical coefficients than the angular position of the phase-matched SHG signal and the measurement presented in Fig. 2.6.

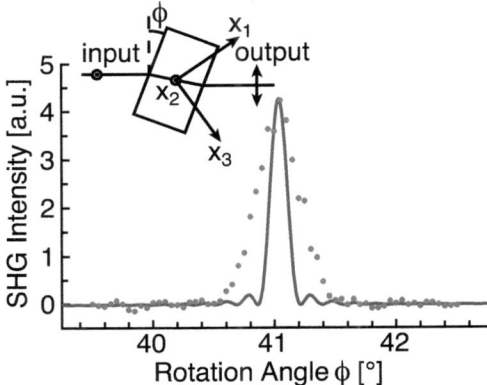

Figure 2.5: The phase matched SHG signal obtained by rotating the DAPSH crystal around the dielectric x_2 axis with a s-polarized impinged fundamental beam at 1.907 µm and p-polarized generated second harmonic light (filled circles). The configuration is schematically illustrated in the inset (top view). The solid theoretical curve has a smaller peak width as the measured signal, because of focusing and/or scattering of the fundamental beam.

For the measurement of d_{111} the fundamental beam was p-polarized and the p-polarized component of the second harmonic light was selected and recorded as a function of the rotation angle (see Fig. 2.6a). A very large number of SHG fringes has been observed in the measurement of d_{111}, due to a very short coherence length in the order of 3 μm for rotation angles where the second harmonic signal is maximal. As a result, the evaluation of d_{111} is very sensitive to small variations of the refractive indices within the error determined by interference measurements (e.g. 2% error in n_1 leads to approximately 20% error in the computation of d_{111}). Therefore, we have first used the measured Maker fringe data for the evaluation of the coherence length and the more precise refractive indices, as described in Appendix A.3. The refractive indices obtained are depicted as diamonds in Fig. 2.3 and agree very well with the observed minima as shown in Fig. 2.6b) (dashed line). They were also used for the evaluation of the nonlinear optical susceptibilities from the Maker fringe experiments (Fig. 2.6a).

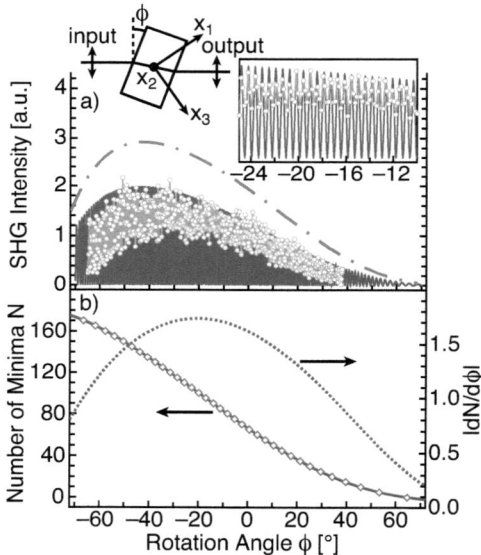

Figure 2.6: a) Maker fringe curve at $\lambda = 1.907$ μm fundamental wavelength obtained by rotating the DAPSH crystal around the dielectric x_2 axis (open circles) for both fundamental and SHG beam p-polarized; a detail is shown in the inset. The upper envelope of the theoretical curve (solid line following the fringes) is in good consistency with the upper envelope of the measured curve. As explained in the text, the actual SHG signal strength is probably given by the sum of the lower and upper envelope of the experimental Maker fringe curve. The upper envelope of the corresponding theoretical Maker fringe curve is shown as dashed-dotted line. b) The number of minima of the Maker fringes N observed in the experiment (open diamonds, only every fifth point is shown for clarity) and the corresponding theoretical curve (solid line). For the number of minima N we chose numbering starting with 0 from the right and included an offset constant in the corresponding theoretical model. The dotted line is the absolute value of the derivative of the number of minima, $|dN/d\phi|$, and is a measure of the density of the Maker fringes.

2.5. Nonlinear optical properties

The upper envelope of the solid theoretical Maker fringe curve in Fig. 2.6a) is in good agreement with the one of the measured data with $d_{111} = 245 \pm 40$ pm/V. The error accounts for the uncertainties in the parameters involved (refractive indices, absorption, sample thickness), as well as the standard deviation from the least-square fitting procedure involving several d_{ijk} coefficients. Best correspondence with the data was obtained for small d_{311} and $d_{113} = d_{131}$ coefficients, which also agrees with the approximate susceptibility tensor in Eq. (2.3). We obtained $d_{311} = 5 \pm 4$ pm/V and $d_{113} = d_{131} = 6.4 \pm 5$ pm/V, where Kleinman symmetry was assumed for the Miller-δ coefficients [90] to relate $d_{113} = d_{131}$ to d_{311}. In the analysis of the Maker fringe measurements an isotropic absorption of 1 cm^{-1} in the ac plane was taken into account for both the fundamental and second harmonic light. The maximum of the theoretical curve (solid line) in Fig. 2.6a), obtained with the given values of d_{111}, d_{311}, d_{113} and d_{131}, will shift toward positive rotation angles by increasing d_{311} and vice versa. The tensor elements d_{133}, d_{333}, d_{313} and d_{331} were not taken into account, since their maximal contribution to the SHG signal is above angles of 55°.

From the SHG measurements presented in Fig. 2.5 and 2.6 we can therefore obtain the following second-order nonlinear optical tensor elements: $d_{111} = 245 \pm 40$ pm/V, $d_{122} = 15 \pm 2$ pm/V, $d_{311} = 5 \pm 4$ pm/V and $d_{113} = d_{131} = 6 \pm 5$ pm/V considering the theoretical curves presented by solid lines. However, these values of the nonlinear optical coefficients represent only a lower limit because of the following reasons. As shown in Fig. 2.6a) the minima of the measured Maker fringes are not zero, which can again be explained by a small distribution of the wave vectors **k** and a very short coherence length. In our opinion the **k**-vector distribution of the focused fundamental beam and due to crystal imperfections leads, on the other hand, also to a reduction of the SHG intensity in a Maker fringe maxima. This is because some part of the beam is always destructively interfering, due to a **k** distribution and the short coherence length. To confirm this assumption we compare the lower envelope of the experimental curve in Fig. 2.6a) with the derivative of the numbered Maker fringe minima with respect to the rotation angle ϕ: $|dN/d\phi|$ in Fig. 2.6b). The latter is a measure of the density of the Maker fringes and is expected to be related to the difference between the experimental and theoretical SHG Maker fringe minima. Figure 2.6 shows, that the lower envelope of the experimental data has a very similar shape as $|dN/d\phi|$. We may further assume, that the reduction of the signal strength in Maker fringe maxima (upper envelope) due to a **k** distribution is equal to the difference between the SHG intensity in Maker fringe minima (lower envelope) and zero (the theoretical value). The dashed-dotted curve in Fig. 2.6a) is the upper envelope of a theoretical Maker fringe curve considering the possible reduction in the experimental signal due to a **k**-vector distribution. The dashed-dotted line was obtained using $d_{111} = 292$ pm/V, $d_{311} = 8$ pm/V and $d_{113} = d_{131} = 10.2$ pm/V, where again Kleinman symmetry was assumed for the Miller-δ coefficients [90]. From this analysis we can conclude that the diagonal nonlinear optical coefficient of DAPSH is very high, $d_{111} = 290 \pm 40$ pm/V at 1.907 µm.

2. Linear and nonlinear optical properties of DAPSH

Finally, we want to compare the measured ratio of d_{111}/d_{122} with the theoretically calculated one according to Eq. (2.2). From the experiments we get the following value $\frac{d_{111}}{d_{122}} = 19 \pm 3$. Taking into account an angle of $\theta_p = 15.5° \pm 0.5°$ (see Fig. 2.1c) between the long axis of the molecules and the polar axis we obtain with the help of Eq. (2.2) $\frac{d_{111}}{d_{122}} = \frac{(\cos\theta_p)^2}{(\sin\theta_p)^2} = 13 \pm 0.9$, which is in reasonable agreement with the experimental result.

Stark spectroscopic and theoretical studies show that the non-resonant hyperpolarizability of the chromophore in DAPSH is increased by 20-50% when compared with that of its DAST analogue [82–84]. For example, respective β_0 values of 90 and $135 \cdot 10^{-30}$ esu for the DAST and DAPSH chromophores were determined by Stark spectroscopy [82], while time-dependent density functional theory gives corresponding values of 163 and $197 \cdot 10^{-30}$ esu [84]. This increase correlates with an enhanced electron-accepting ability of the pyridinium unit that is also evidenced by visible absorption, NMR and electrochemical data. A larger value of the macroscopic nonlinear optical susceptibility of DAPSH compared to DAST is also expected due to the about 6% larger chromophore density N, a higher projection factor $\cos^3\theta_p$ of about 8% and a local field factor $f_1^{2\omega}(f_1^\omega)^2$ that is almost 50% larger as compared to DAST. On the other hand, the hyperpolarizability of densely packed chromophores in a crystal may be reduced due to molecule-molecule interactions [91]. Also theoretical calculations have shown [84] that the first hyperpolarizability of stilbazolium dyes strongly depends on the molecular environment. Nevertheless, even considering a lower limit of d_{111} of 245 ± 40 pm/V at 1.907 μm, the nonlinear optical coefficient is about 15% larger than the one in DAST and is, to the best of our knowledge, the largest non-resonant nonlinear optical coefficient measured to date.

2.6 Conclusion

We have determined the absorption properties and the principal refractive indices of the organic nonlinear optical stilbazolium salt DAPSH in the wavelength range from 0.7 μm to 2 μm. A large anisotropy of the refractive index of $\Delta n = 1.17 \pm 0.06$ at $\lambda = 0.83$ μm and $\Delta n = 0.83 \pm 0.04$ at $\lambda = 1.55$ μm has been measured, which is explained by the beneficial alignment of the nonlinear optical chromophores.

Furthermore, we have measured the nonlinear optical coefficients for second harmonic generation d_{111} and d_{122} at 1.907 μm. DAPSH has a very high diagonal second-order nonlinear optical tensor element of at least $d_{111} = 245 \pm 40$ pm/V, which is higher than that of the well-investigated DAST ($d_{111} = 210 \pm 55$ pm/V at $\lambda = 1.907$ μm [66]). We believe that the measured SHG intensity has been reduced due to a very short coherence length, a k-vector distribution caused by focusing the beam and crystal imperfections, and therefore even higher d-values of $d_{111} = 290 \pm 40$ pm/V are reasonable. Hence, this material is a very interesting candidate for high-speed electro optic modulation and THz generation.

CHAPTER 3

Electro-optic single-crystalline organic waveguides and nanowires grown from the melt*

Organic nonlinear optical materials have proven to possess high and extremely fast nonlinearities compared to conventional inorganic crystals, allowing for sub-1-V driving voltages and modulation bandwidths of over 100 GHz. Compared to more widely studied poled electro-optic polymers, organic electro-optic crystals exhibit orders of magnitude better thermal and photochemical stability. The lack of available structuring techniques for organic crystals has been the major drawback for exploring their potential for photonic structures. Here we present a new approach to fabricate high-quality electro-optic single crystal waveguides and nanowires of configurationally locked polyene DAT2 (2-(3-(2-(4-dimethylaminophenyl)vinyl)-5,5-dimethylcyclohex-2-enylidene)malononitrile). The high-index-contrast waveguides ($\Delta n = 0.54 \pm 0.04$) are grown from the melt between two anodically bonded borosilicate glass wafers, which are structured and equipped with electrodes prior to bonding. Electro-optic phase modulation is demonstrated for the first time in the non-centrosymmetric DAT2 single crystalline channel waveguides at a wavelength of 1.55 µm. We also show that this technique in combination with DAT2 material allows for the fabrication of single-crystalline nanostructures inside large-area devices with crystal thicknesses below 30 nm and lengths of above 7 mm.

3.1 Introduction

Organic materials have been proposed for electro-optic applications, especially for fast optoelectronic switching and modulation, because of their high electro-optic coefficients and low dielectric constants resulting in extremely fast response times compared to presently employed inorganic materials [16, 17, 19, 21, 23, 28, 36, 62, 92, 93]. For these applications it is particularly crucial to achieve a stable noncentrosymmetric packing of the nonlinear optical chromophores. There are two main approaches for acentric molecular orientation: poled electro-optic polymers

---

*This chapter, together with Appendix B has been published in Optics Express **16** (15), 11310–11327 (2008).

3. Electro-optic single crystal waveguides of DAT2

and single crystals. Polymers are often easier to process in thin films compared to their crystalline counterparts and offer a relatively high fabrication flexibility. On the other hand polymers with electro-optic molecules have to be poled under high external fields to achieve non-centrosymmetry and their long-term orientational stability is limited by the relaxation rate of the chromophores, especially in micro- and nanostructured devices. Therefore, one way to overcome the limitations of poled polymers is to use organic nonlinear optical crystalline materials, which offer a high density chromophore packing and stable chromophore orientation. Additionally, organic nonlinear optical crystals present several orders of magnitude better photostability than the polymers [94].

Although the inherently superior organic nonlinear optical crystalline materials are very promising for high-speed photonic applications, their very large scale integration has not yet been possible due to many unsolved problems in the production and processing of single crystals. A substantial progress was partially achieved by the development of crystalline thin films, whose planar structure is evidently more compatible with waveguide configurations compared to bulk crystals [95, 96]. Nevertheless the thickness control of such films is often limited and the application of standard photolithography to planar 2D crystalline films generally remains a challenging task, due to the incompatibility of the solvents used for conventional optical lithography with the organic crystals [63]. In the past, most of the techniques to fabricate organic electro-optic crystalline waveguides were concentrated on the ionic salt DAST (4-N, N-dimethylamino-4'-N'-methyl-stilbazolium tosylate) due to its high electro-optic figure of merit at 1.55 µm, i.e. $n^3 r = 455$ pm/V, combined with a low dielectric constant of $\epsilon = 5.2$, resulting in an about one order of magnitude better device-relevant parameter $n^3 r/\epsilon$ compared to the benchmark $LiNbO_3$ [63]. Several microstructuring techniques were developed alternative to conventional photolithography such as photobleaching [70, 97], femtosecond laser ablation [98], ion implantation [99] and direct e-beam structuring [73]. A very interesting technique based on graphoepitaxial melt growth of DAST was proposed by Geis et al. [71], however the applicability of this method is limited due to an insufficient thermal stability of DAST at its melting temperature. Therefore rapid growth rates were necessary to prevent decomposition of the DAST chromophores and consequently only moderate crystal quality was obtained. Furthermore, melt growth on a structured substrate may lead to an irregular upper surface and thickness of the crystal.

Here we use a recently developed organic nonlinear optical material which is thermally stable above the melting temperature. We were able to produce crystalline organic waveguides with a new fabrication technique, in which the material is grown from the melt in the desired geometry and shape. The crystalline waveguides were produced with an approach derived from the widely investigated technique of capillary methods, where in general the melt or solution is brought in-between two plates by capillary action and subsequent slow cooling or evaporation respectively, forces the molecules to crystallize [100, 101]. It has already been shown that the growth of high quality single crystalline thin films of DAT2 (2-(3-(2-(4-dimethylaminophenyl)vinyl)-5,5-

dimethylcyclohex-2-enylidene)malononitrile) from the melt, vapor and from the solution is very promising [60, 102, 103]. However, in all these methods the growth direction, thickness and the growth position of the rectangular crystals could not be controlled. One possible attempt to overcome these limitations is to let the melt or the solution flow into predefined channels by capillary force and let the material crystallize there. We show that this is possible with DAT2, also for channel dimensions down to only 25 nm height and a few μm width. This is a surprising result, since most of the materials, especially highly nonlinear optical chromophores, will experience a limited flow, crystallization and single-crystallinity in submicrometer size structures. With the method reported here and the DAT2 molecule developed in our laboratory for this application, the grown single crystalline structures possess both high thermal and photochemical stability and good mechanical protection.

3.2 Material

The configurationally locked polyene DAT2 chromophore consists of a phenylhexatriene bridge between a dimethylamino electron donor group and a dicyanomethylidene electron acceptor, leading to a first hyperpolarizability $\beta = 1100 \cdot 10^{-40}$ m^4/V at 1.9 μm that is among the largest microscopic nonlinearities of chromophores that crystalize in a noncentrosymmetric crystalline arrangement [60]. DAT2 crystals exhibit monoclinic point group symmetry 2 with two chromophores in the unit crystal cell. The lattice parameters are $a = 6.130$ Å, $b = 7.424$ Å and $c = 20.258$ Å with $\alpha = \gamma = 90°$ and $\beta = 96.75°$ [60]. The orientation of the molecules with respect to the unit cell in the ac plane is shown in Fig. 3.1.

DAT2 crystals are particularly interesting since they show a large second harmonic generation signal in the powder test of about two orders of magnitude greater than that of urea [60]. Compared to the well-studied DAST crystals, DAT2 exhibits a considerably enhanced temperature stability with the weight-loss temperature at $T_i = 293°$C (involving sublimation and/or decomposition) and a melting temperature of $T_m = 235°$C, as well as a large temperature difference of 80°C between the melting temperature and the recrystallization temperature. This is very promising for the crystal growth from the melt. Most of the other high-nonlinearity molecules decompose before the melting temperature and therefore only solution growth techniques can be used [28]. DAT2 also shows a very high tendency to form thin films; the growth of high-quality single-crystalline thin films from vapor [102] and solution [103] have been recently demonstrated.

3.3 Sample preparation

In this section the technique of fabricating crystalline waveguides of DAT2 is described. The first step is to pattern the waveguide geometry in a substrate material and to deposit electrodes. In a second step, a cover is anodically bonded to the substrate to delimit the volume for the crystallization of the electro-optic active material DAT2. Finally, DAT2 powder is placed on

3. Electro-optic single crystal waveguides of DAT2

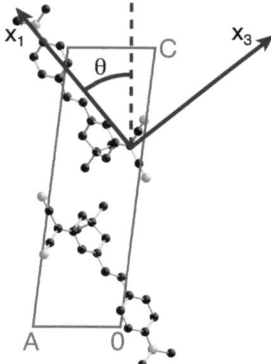

Figure 3.1: A view of the unit cell along the polar crystallographic b axis (two fold symmetry axis) showing the alignment of the chromophores in the ac plane with respect to the unit cell; θ is the angle between the dielectric axis x_1 and the normal to the bc plane (dashed line) and was determined as described in Appendix B. The good agreement of the intramolecular donor-acceptor direction with the experimentally determined orientation of x_1 can be seen. The hydrogen atoms are omitted for clarity.

the substrate at the edge of the cover and melted such that the melt can flow through the waveguide channels into the structure and crystallize there.

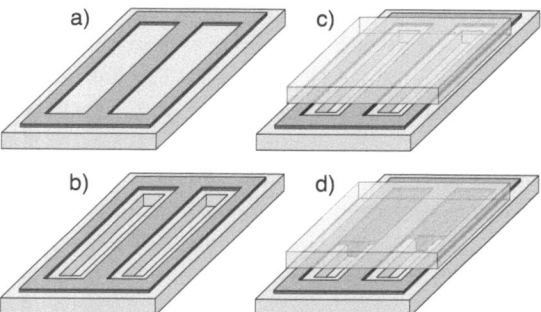

Figure 3.2: Processing steps for the fabrication of electrode equipped waveguide channels. a) 40 nm chromium and 50 nm amorphous silicon were deposited and structured on a borosilicate wafer by standard photolithography. b) The straight waveguide structure was patterned in-between neighboring chromium/silicon stripes by reactive ion etching. c) A cover borosilicate glass was anodically bonded to the fabricated structure to delimit the waveguide volume in vertical direction. d) The cover glass was shorter than the waveguide length, such that the fabricated channels were accessible for the melt of DAT2 and the material could flow in by capillary action and crystallize there.

3.3. Sample preparation

To manufacture the structured and electroded substrate two photolithographic steps were required. In a first step 40 nm chromium and 50 nm amorphous silicon were deposited on a borosilicate glass wafer by standard photolithographic processing, and a standard lift-off technique (see Fig. 3.2(a)). The chromium was used as electrical contacts to the waveguide in the final structure and silicon was required for anodic bonding. In the second lithographic step grooves with dimensions of 1.5-5 µm width were structured in-between the electrodes into the commercial photoresist AZ6632 and then transferred to the borosilicate substrate wafer by subsequent reactive ion etching (RIE) (see Fig. 3.2(b)).

A cover borosilicate glass shorter than the length of the patterned grooves was anodically bonded to the structured substrate wafer, such that the left and right ends of the grooves in the substrate protruded from the edges of the cover glass and the DAT2 melt could flow in (see Fig. 3.2(c)). Anodic bonding is one of a number of techniques used in the semiconductor industry for wafer bonding [104–106]. It is well-established and was in 2004 reported to account for the majority of all packaging applications for microelectromechanical (MEMS) devices [106]. The substrate and cover material is a commercial standard borosilicate glass suitable for anodic bonding to silicon due to a thermal expansion coefficient that matches the one of silicon in order to avoid thermal stress. The borosilicate wafers had a thickness of 200 ± 25 µm, an average roughness of the surface of less than 1.5 nm and a total thickness variation of less than 10 µm. The anode of a high voltage power supply was contacted to the deposited chromium and the cathode to a wolfram plate pressed slightly against the cover glass. The bonding process was performed at a temperature of 456°C by applying a dc voltage of 700 V to the two electrodes. The silicon deposited on top of chromium acts as a sodium diffusion barrier. It has been shown that in similar geometries also very thin layers of 20 nm silicon are sufficient as sodium diffusion barriers [107].

Finally a few milligrams of DAT2 powder were placed onto the substrate immediately at the edge of the cover glass, where the structured grooves in the substrate glass were still present and accessible for the melt. To one side of the sample a 100 mm long and 20 mm wide aluminium foil was attached by silver paint. The sample was put into a glass tube, which was filled with argon at atmospheric pressure and sealed with a vacuum valve. The sample was then placed in the middle of a heating coil and heated until the material started to melt and to flow into the channels (see Fig. 3.2(d)). Instead of moving the sample into a cooling section as it is usually the case for Bridgman methods [28], we have chosen cooling ramps on the order of 30°C per hour. The aluminium stripe attached to one side of the sample should ensure an asymmetry in the temperature profile along the channels to prevent several nucleation points. The total growth time was less than eight hours. This relatively short growth time is another advantage compared to e.g. solution grown thin films that typically require 2-10 weeks growth time [103].

3.4 Crystal orientation and crystal quality

The orientation of the melt grown thin film crystals was investigated by X-ray diffraction $\theta - 2\theta$ scans. Note that the fabrication steps were slightly different for the sample used for this measurement, since anodic bonding prevents to access the crystal by the X-rays. Instead of the bonding process, the cover borosilicate glass was mechanically pressed to the substrate during the growth and removed afterwards. The X-ray measurements were performed with a STOE Stadi P diffractometer in reflection mode (CuKα_1 radiation, $\lambda = 1.54056$ Å). We observed Bragg reflections assignable to a family of (00h) planes coplanar to the substrate surface (see Fig. 3.3). The measured reflections correspond to a real-space periodicity of 20.1 ± 0.3 Å. Hence, melt grown structures have the same crystal orientation as the solution grown thin films, which means that the crystal surface parallel to the substrate is the ab-crystallographic plane [103].

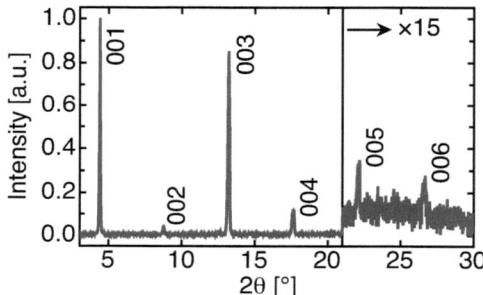

Figure 3.3: X-ray diffraction $\theta - 2\theta$ scan of a melt grown thin film crystal of DAT2. The reflections correspond to a single lattice constant normal to the surface of 20.1 ± 0.3 Å.

By rotating the grown waveguide structures between crossed polarizers in a microscope, the orientation of the a and b-axes with respect to the grooves could be determined, since the crystals became homogeneously dark for a polarization direction parallel or perpendicular to the dielectric x_2 axis, which is set parallel to the crystallographic b-axis in the point group symmetry 2. However, the a and b-axes are not easy to distinguish from each other. As detailed in Appendix B, the directions of the crystallographic a and b axes for the solution grown thin films were determined by polarized second-harmonic generation measurements, where it was found that the a axis is parallel to the long axis of the rectangular crystals. From a birefringence measurement with the help of a tilting compensator B (Zeiss) and the measured refractive indices (see Appendix B) we could distinguish the crystallographic a- and b-axes also for the melt grown crystals. DAT2 crystallizes with the polar b axis perpendicular to the waveguide channels, which is essential for applying the electro-optic effect (see Fig. 3.4). For most of the lower-nonlinearity molecules with shorter conjugation length, for which melt growth in fibers was previously demonstrated (BNA, COANP, FNPH, mDHB, mDNB, mNA, MNA, NPAN,

PNP and others) this is not the case (polar axis is parallel or almost parallel to the waveguide axis) with a few exceptions (DAN, MAP, NPP and others) [28, 108].

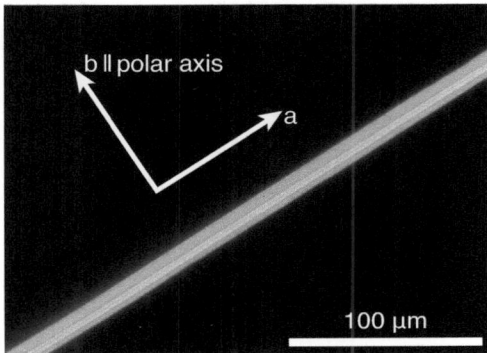

Figure 3.4: Transmission microscope image of a DAT2 crystal grown from the melt in a groove of a correspondingly structured substrate. The crystallographic a- and b-axes are parallel and perpendicular to the waveguide respectively.

The quality of the melt grown crystals was investigated by optical microscopy and the corresponding end-facets after cleaving by SEM. After cooling of samples with a larger cross-section, cracks can be induced, which are then mainly parallel to the crystallographic a-axis. In most of our cases the a-axis is parallel to the direction of the waveguide and thus usually only one crack is present but can be seen almost along the entire waveguide (see Fig. 3.4). The characteristics of the crack indicate that the crystal orientation in the waveguide channel is perfectly preserved for several millimeters as also no deviations could be noticed under the microscope between crossed polarizers. SEM inspection of such cracks indicated that they are flat and regular, therefore the waveguide mode scattering may be of minor importance compared to the side-wall roughness. Figure 3.5 shows a scanning electron microscopy (SEM) image of a cleaved straight waveguide structure with a thickness of $h = 0.97 \pm 0.12$ µm (see Fig. 3.5). After cleaving the sample perpendicular to the waveguides, the end-facets of the melt grown crystals appear in general sharp and flat (see Fig. 3.5). Additional polishing was therefore not required and waveguiding was clearly observed by using conventional end-fire coupling at the telecommunication wavelength of 1.55 µm.

3.5 Refractive indices and electro-optic modulation

The refractive indices of DAT2 were measured with three independent measurement methods. The details of the measurements are described in Appendix B.

DAT2 crystalline waveguides of good optical quality grown from the melt in a waveguide channel equipped with electrodes were used to demonstrate electro-optic phase modulation. The measurements were performed with infrared light at $\lambda = 1.55$ µm and the electro-optic

3. Electro-optic single crystal waveguides of DAT2

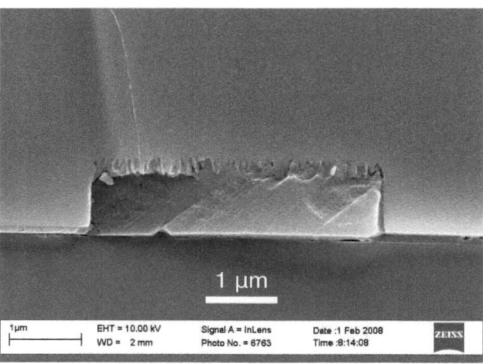

Figure 3.5: Scanning electron micrograph of an end-facet of a melt grown DAT2 crystal.

properties were determined by applying an ac field to the electrodes. To detect phase modulation the sample was incorporated into one arm of a Mach-Zehnder interferometer. The signal beam was end-fire coupled into the waveguide using a 50× microscope objective. The output light of the waveguide was imaged with a 40× objective and combined with the reference beam to create interference. When a voltage was applied to the electrodes, the phase shift created by the electro-optic effect in the crystal causes an intensity modulation when the signal and the reference beam interfered at the exit of the interferometer. The reference beam passed an additional compensator, which allowed to set the working point of the modulator. The distance between the electrodes was $d = 13$ µm, with the waveguide in the middle. The characterization was done with a sinusoidal voltage with peak amplitude of $U = 10$ V applied to the electrodes. The resulting linear amplitude modulation was measured with a lock-in amplifier (Stanford Research Systems SR830). The waveguides with dimensions 1-1.6 µm (height) and 3.5-7 µm (width) were not single mode. However, by applying an electric field to the electrodes, optical modulation has been demonstrated.

Linear electro-optic effects are most commonly described by considering field-induced changes of the optical indicatrix in noncentrosymmetric materials and can be described by the tensor r_{ijk}:

$$\Delta\left(\frac{1}{n^2}\right)_{ij} = r_{ijk} E_k \ , \tag{3.1}$$

where r_{ijk} is given in the Cartesian dielectric system as shown in Fig. 3.1 and where summation over common indices is assumed. From the orientation of the chromophores we expect that r_{112} and $r_{211} = r_{121}$ are the largest electro-optic tensor elements of DAT2. Therefore the electro-optic modulation measurements were performed with a TM mode propagating along the crystallographic a-axis. Due to the inclined indicatrix in ac-plane the phase shift for TM modes in the waveguide is not only caused by r_{112} but also by r_{332} and $r_{132} = r_{312}$. Theoretical hyperpolarizability studies have shown that the effective hyperpolarizability β_{112} in the reference

3.5. Refractive indices and electro-optic modulation

frame of a diagonal dispersionless dielectric tensor is more than 60 times larger than β_{132}, β_{332} [109]. Hence we may neglect the change of the refractive index of the waveguide caused by r_{132} and r_{332} and only take into account the projection factors of the dielectric system in which r_{ijk} is given onto the reference frame of a TM mode based on standard tensor transformation. The inclination angle θ was determined by refractive index measurements and conoscopy, as detailed in Appendix B.

The investigated electro-optic coefficient r_{112} is therefore given by

$$r_{112} = \frac{\delta I_M}{\Delta I_M} \cdot \frac{1}{(\cos\theta)^2} \cdot \frac{2\lambda d}{\pi l n_{\text{eff}} n^2 \delta U} , \qquad (3.2)$$

where ΔI_M is the measured intensity difference between fully destructive and constructive interference at the photodetector, δI_M is the intensity modulation detected in the working point due to the field modulation $\delta E = \delta U/d$ in the waveguide, where we assumed in a first approximation equal permittivity constants for the electro-optic active material and the surrounding borosilicate glass, θ is the angle between the dielectric axis x_1 and the normal to the crystallographic ab-plane, l is the length of the crystal in the propagation direction of the light, n_{eff} is the effective refractive index experienced by the waveguide mode and n is the bulk refractive index of the chosen polarization. In Eq. (3.2) both n and n_{eff} appear since in a first order approximation the same proportionality constant was used to relate n_{eff} and n as to relate the change of the corresponding refractive indices due to the electro-optic effect, i.e. Δn_{eff} and Δn.

The electro-optic coefficient has been determined for two single crystal waveguides with different cross section. The effective refractive index for the two configurations was calculated with the full vectorial complex FD Generic mode solver of the commercially available software OlympIOs [110]. Using Eq. (3.2) the lower limit of the electro-optic coefficient $r_{112} = 7\pm 1$ pm/V was obtained at a wavelength of 1.55 μm and the values for the two examined crystal waveguides were within experimental error in good agreement. This value presents a lower limit since the waveguides were not single mode and since a possible reduction of the field inside the waveguide due to a lower permittivity cladding was neglected. Note that the values of the electro-optic coefficients in bulk DAT2 crystals are not available yet.

Preliminary results show that the investigated waveguides had propagation losses of around 14 dB/cm, which were estimated by measuring the optical power with and without the sample in the beam path and taking into account Fresnel reflections at the two end-facets and a mode coupling efficiency of 80%. Since the intrinsic material absorption losses are expected to be below 4 dB/cm (absorption constant $\alpha < 1$ cm^{-1}), the propagation losses may be mainly caused by the substantial top-wall roughness induced by the etch process (see Fig. 3.5), affecting especially higher-order modes. Several studies devoted to ion-enhanced chemical etching of borosilicate glass in CHF_3/CF_4 plasma have addressed the problem of surface and sidewall roughnesses [111, 112]. The resulting rough surfaces from dry etching of borosilicate glasses are caused by the formation of nonvolatile halogen compound products generated in the etching process. The roughnesses of etched surfaces can be greatly improved by physical sputter-etching to remove nonvolatile products and minimize associated micromasking more efficiently, using for example inductively coupled plasma reactive ion etching [111, 113, 114]. The side-wall roughness obtained by conventional RIE, combined with the high refractive index contrast $\Delta n = n_{\text{TM}} - n_{\text{glass}} = 0.54 \pm 0.04$ experienced by the TM modes between the DAT2 material and the surrounding borosilicate glass ($n_{\text{glass}} = 1.456$) at a wavelength of 1.55 µm may cause a substantial scattering loss. We therefore believe that an optimized etch process to smoothen the side-walls will strongly reduce the propagation losses. The high refractive index contrast is however of particular importance to attain a high degree of optical mode confinement in very large scale integrated circuits, such as in microring resonators and in photonic bandgap structures.

3.6 Growth of single crystalline nanowires and nanosheets (sub 30 nm thick crystals)

It was recently reported that slotted silicon waveguide configurations allow for a considerable improvement of the tunability of electro-optic modulators, sensors and all-optical switching devices over the current state of the art [22, 24]. The essential feature of a slot waveguide is that two high-index silicon stripes are separated by a small distance on the order of 20-140 nm filled with a lower-index material. The optical field intensity in such a structure tends to concentrate within the low-index slot for the polarization perpendicular to the slot. Therefore and also due to the close proximity of the electrical contacts in a slotted geometry filled with an electro-optic active material, a very large refractive index change for a given modulation voltage can be obtained compared to a more conventional waveguide with external electrodes. It was shown that even at very small gaps, on the order of 20-40 nm, the modulation bandwidth per Volt still increases for narrower and narrower gaps [24]. The method presented here, offers the possibility to grow single crystalline nanowires inside nano-sloted structures with slot dimension of at least down to 25 nm.

Nanometer size channels were realized with a simple one-layer approach and a standard lift-off process. The commercial photoresist ma-N 415 was spin coated on the borosilicate substrate. A standard UV lithography process was performed to generate the required ridges in the photoresist with widths between 3 and 11 µm. Approximately 20 nm silicon was then deposited preserving the undercut for subsequent lift-off. Finally, to obtain the 25 nm high and several µm wide channels, the fabricated pattern in silicon was terminated by anodic bonding of a second borosilicate wafer to the structure as described in Section 3.3. The growth method was also the same as in Section 3.3. No attempts were made to manufacture even thinner channel heights yet, since from a fabrication and detection point of view we achieved our limits. A microscope image and a scanning electron micrograph of the grown crystals are shown in Fig. 3.6 and 3.7 respectively. Amazingly, a single crystal thickness of only 25 nm was realized

3.6. Growth of single crystalline nanowires and nanosheets

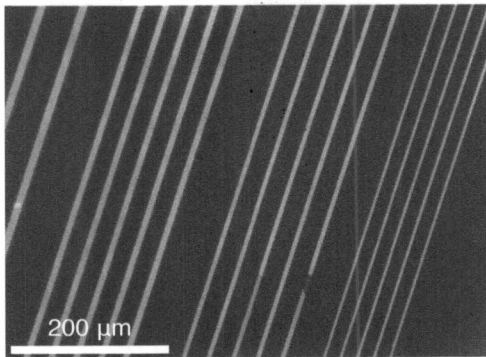

Figure 3.6: Transmission microscope image of approximately 25 nm thick DAT2 crystalline stripes grown from the melt as seen between crossed polarizers. The corresponding end-facet is shown in Fig. 3.7.

perpendicular to the *ab*-crystallographic plane. This corresponds to only approximately a dozen unit cells in this direction.

Furthermore, the lateral crystalline dimension of the reported nanowires could be increased by two orders of magnitude and therefore the growth of ultra-thin large-area crystals has been realized. The processing for the fabrication of nanosheets was analogous to the one depicted in Fig. 3.2, except that the chromium-silicon bonding areas were structured much further apart, such that channel dimensions of about 90 nm height and 0.5 mm width were formed. Inspection of the crystalline nanosheets under the microscope between crossed polarizers showed that large single crystalline areas are obtained with domain sizes on the order of 10'000 µm^2 where the crystal orientation is preserved. The nanosheets grown in our first experiments are single crystalline on most of the area but show small holes (black spots in Fig. 3.8).

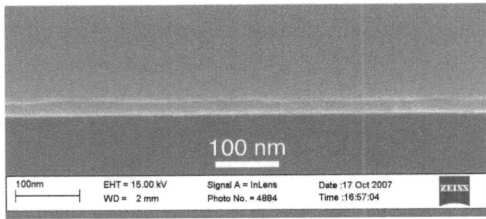

Figure 3.7: Scanning electron micrograph of an end-facet of an approximately 25 nm thick DAT2 crystal grown from the melt.

These nanosheets are highly interesting also because with their help DAT2 crystals can be grown in closed waveguide configurations, such as microrings (see Fig. 3.8). Waveguide structures of any shape can be fabricated by RIE in the substrate borosilicate wafer (analogously to the straight waveguide structuring in Fig. 3.2) and they can be filled by flowing the melt of

DAT2 through the nanosize channels to the fabricated structure. The orientation of the crystal in the waveguide-structure is determined by the corresponding single crystalline domain of the nanosheet.

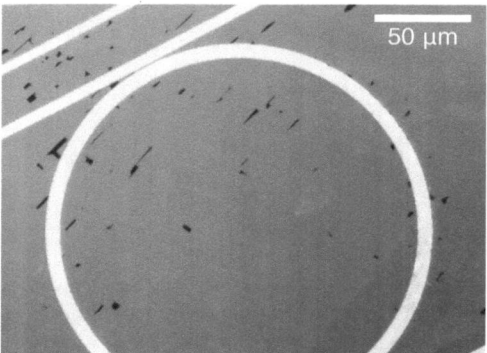

Figure 3.8: Transmission microscope image of an approximately 90 nm thick crystalline nanosheet of DAT2 grown from the melt as seen between crossed polarizers. The black spots indicate regions without crystalline material. Two crystalline domains can be seen as noticeable in the lower right corner, where the crystal orientation rotates by $11 \pm 1°$. Grooves of any shape structured in the substrate borosilicate glass can be efficiently filled by flowing the melt of DAT2 through nanosize channels to the structure, a 0.5 µm deep microring resonator structure in this case. The orientation of the crystal in the ring-structure is the same as throughout the entire single crystalline domain.

3.7 Conclusion

In conclusion, we developed a new approach to fabricate high-quality single-crystalline high-index-contrast waveguiding structures and demonstrated electro-optic modulation at the telecommunication wavelength 1.55 µm in DAT2 waveguides grown from the melt. Standard photolithography, plasma-etching and anodic bonding technique were used for the electrode equipped microchannel fabrication. The crystals grown inside the channels by a simplified variation of the Bridgman method were efficiently protected against chemical and mechanical damage. Due to the good optical quality of the crystals they can be used as optical waveguides. The estimated optical losses of approximately 14 dB/cm in the first experiments reported were mostly due to scattering caused by a side-wall roughness originating from plasma etching and the material combination DAT2/borosilicate glass with a high refractive index contrast $\Delta n = 0.54 \pm 0.04$. This scattering is expected to be greatly reduced by using inductively coupled plasma RIE and single mode structures. First electro-optic phase modulation experiments could be demonstrated using DAT2 channel waveguides at telecommunication wavelengths. The electro-optic coefficient of DAT2 determined using these waveguide modulators is at least

3.7. Conclusion

$r_{112} = 7 \pm 1$ pm/V. The exceptional crystal growth possibilities such as the growth of sub 30 nm thick crystalline nanowires and nanosheets from the melt as well as the high thermal and photochemical stability make DAT2 a particularly interesting material. Beside the relatively fast growth time and low cost production, no photolithographic step has to be directly applied to the organic crystals. The technique is also so versatile that it can be applied to several other organic nonlinear optical materials as long as the growth of them from the melt is feasible. Further effort is underway [109, 115] to combine the favorable crystal growth properties of DAT2 with a more optimal chromophore alignment in the crystal to achieve higher nonlinear optical coefficients comparable to the ones of the ionic salts, such as DAST and the presently best material DAPSH [116].

3. Electro-optic single crystal waveguides of DAT2

CHAPTER 4

Electro-optic tuning and modulation of single-crystalline organic microring resonators*

We present, for the first time to our knowledge, the fabrication and electro-optic tuning of single-crystalline organic microring resonators. In recent years, optical microring resonators have proven to be highly suitable building blocks for the realization of very large scale integrated photonic circuits. In particular, microresonators based on organic materials are very promising for ultra-fast electro-optic applications, due to the electronic nature of the electro-optic response preserving the modulation performances beyond 100 GHz. However, the implementation of organic materials into real devices is presently limited by intrinsic properties such as thermal and photochemical stability. In contrast to polymer waveguiding structures realized previously, our crystalline thin film devices feature an excellent long-term stability of the chromophore orientation, superior photochemical stability, and do not require high-field poling prior to operation. The introduced thin film fabrication method significantly reduces fabrication complexity of organic crystalline electro-optic waveguides, compared to previously developed techniques. We have fabricated crystalline COANP (2-cyclo-octylamino-5-nitropyridine) microring resonators with resonance contrast up to 10 dB, ring waveguide propagation losses of about 10 dB/cm, a free spectral range of 1.6 nm, a finesse of up to 20 and a corresponding Q-factor of about 20'000, measured in the telecom wavelength range around 1.55 µm. We have demonstrated resonance wavelength tuning at the rate of 0.13 GHz/V (1.1 pm/V).

4.1 Introduction

With increasing penetration of optical data transmission from long-haul connections into local area networks or even chip-to-chip and intra-chip connections, the role of large-scale integrated optics becomes more and more important. In order to integrate filtering and switching devices on

---

*This chapter, without Appendix C has been published in the Journal of the Optical Society of America B: Optical Physics **26**(5), 1103–1110 (2009)

a wafer scale, whispering gallery mode microresonators represent one of the most compact and efficient solutions. In particular, compared to the presently fastest Mach-Zehnder modulators in long-haul transmission systems, which are made mostly by using an expensive inorganic bulk crystalline technology based on the ferroelectric material LiNbO$_3$, microring resonators have evidently the potential of higher miniaturization.

Recently, the emerging field of silicon based photonics has led to several impressive results regarding the miniaturization of photonic circuits [12, 13]. Especially the matured processes, the access of a huge knowledge base, the ability to pattern nanoscale structures as well as the low optical loss and high mode confinement in silicon waveguides make silicon-on-insulator an attractive material system for integrated photonics. On the other hand, silicon based electro-optic devices rely on the relatively weak free-carrier dispersion effect, which typically requires a high driving power for obtaining a significant modulation depth. The fastest free-carrier injection broadband light modulation in silicon technology allowed modulation up to 30 GHz [14]; ultra-fast optical modulators have not been reported mainly because of silicon's weak ultra-fast nonlinearity. Higher modulation speeds can be achieved by using the intrinsic electro-optic effect such as in conventional LiNbO$_3$ modulators, where standard commercial products support data rates up to 40 Gb/s. However, owing to the large dispersion of the dielectric constant from the modulation to optical frequencies, there are strong doubts about the feasibility of LiNbO$_3$ based electro-optic Mach-Zehnder modulators with a bandwidth above 100 GHz [19, 20]. On the other hand, organic materials are ideally suited for high-speed modulators due to the ultra-fast and almost pure electronic origin of the optical nonlinearity. For example, high-speed electro-optic modulators based on organic polymers with high bandwidth of up to 165 GHz and extremely low switching voltages below 1 V have been demonstrated [15–17]. Also very high modulation frequencies are expected by combining the highly nonlinear characteristics and the ultra-fast electro-optic effect of organic materials with good high-index guiding properties of silicon-on-insulator devices [23, 117]. These considerations indicate that photonic devices based on organic materials could play a major role in the development and improvement of future optical components in highly integrated wavelength-division multiplexing systems.

The organic electro-optic materials investigated for their use in photonic applications can be mainly divided in two classes: poled polymers and organic crystals. Polymers are in general malleable and allow to exploit conventional device production techniques common to semiconductor industry, as photo- and electron-beam lithography along with dry and wet etching processes, but also more specific procedures like replication by micro-embossing. On the other hand, polymers with electro-optic molecules have to be poled under high external fields to achieve non-centrosymmetry and their long-term orientational stability is limited by the relaxation rate of the chromophores, especially in micro- and nanostructured devices. One way to overcome the limitations of poled polymers is to use organic electro-optic crystalline materials, which offer a high-density chromophore packing and stable chromophore orientation. Additionally,

4.1. Introduction

organic electro-optic crystals present several orders of magnitude better photostability than polymers [94].

The relatively high fabrication flexibility of electro-optic polymers comes into notice by the fact that several photonic device structures even with submicron scale features have been successfully realized. For example organic electro-optic polymers have been structured by state of the art electron-beam lithography combined with plasma reactive etching techniques [118] to form photonic crystal slabs, furthermore electro-optic microring resonators have been patterned by micro-embossing [36], direct electron-beam writing [37] or direct photodefinition [35, 119]. On the contrary the very large scale integration of the inherently superior organic electro-optic crystalline materials in high-speed photonic applications has not yet been possible due to many unsolved problems in the production and processing of single crystals.

In the past, many techniques have been investigated for the fabrication of organic electro-optic crystalline devices such as femtosecond laser ablation [98], photobleaching [70, 97], graphoepitaxial growth [71], ion implantation [99], direct electron-beam structuring [73], and modified photolithography [120]. Despite these promising fabrication techniques, organic single crystal growth as well as structuring with adequate quality and precision for optical applications has remained challenging. Probably these difficulties account for the fact why only relatively simple integrated devices such as waveguide phase modulators or Mach-Zehnder modulators are reported in literature.

We recently developed a new fabrication technique in which the melt of the organic material flows into predefined channels by capillary force and crystallizes there upon cooling. By this method it has already been shown that the growth of high-quality single-crystalline phase modulators from the melt is possible [121].[1]

In this paper we demonstrate, to the best of our knowledge, the first electro-optic modulation and tuning of a single-crystalline organic microring resonator. The crystalline electro-optic waveguide structures were grown from the melt in prefabricated grooves patterned in standard borosilicate wafers. The power of the fabrication technique is not only illustrated by the achieved results below but also by a number of non-obvious key benefits such as a high photochemical stability of the grown single-crystalline structures, a good mechanical protection of the generally brittle organic crystals, a very accurate control of geometrical device parameters since the entire photolithographic structuring is performed on high-quality inorganic substrates, consequently no photolithographic step has to be directly applied to the organic crystals, and probably most importantly the technique is so versatile that it can be applied to different organic electro-optic materials as long as the growth of them from the melt is feasible.

---

[1]In Appendix C, electro-optic amplitude modulators based on integrated Mach-Zehnder interferometers with single-crystalline DAT2 (2-(3-(2-(4-dimethylaminophenyl)vinyl)-5,5-dimethylcyclohex-2-enylidene)malononitrile) as active material are reported, which have been obtained by using a very similar fabrication concept as presented in this chapter.

4.2 Material and crystal structure

COANP (2-cyclo-octylamino-5-nitropyridine) is an amphiphilic molecule, where the neutral part is formed by an octyl ring and the polar part by the nitropyridine [122, 123]. It has been shown that the growth of large area crystalline thin films of COANP with a high surface quality from the melt is relatively easy [101], which can partially be explained by the relatively low melting point of 70.9°C leading to a small thermally induced strain in the grown crystals. The films that have been reported in the literature were roughly 2 cm² in area with thicknesses of a few to several dozen micrometers, where the crystallographic a axis was perpendicular to the largest crystal surface. The crystals exhibit orthorhombic point-group symmetry $mm2$, therefore the three optical main axes of the index ellipsoid are oriented along the crystallographic axes. We have chosen the organic material COANP because it is a widely investigated material, where high-quality thin film crystals are readily obtainable with the polar axis (crystallographic c-axis) oriented parallel to the thin film plane, which is of utmost importance considering the device configuration as described below. Nevertheless, the electro-optic coefficients reported for COANP show moderate values ($r_{33} = 15 \pm 2$ pm/V at 632.8 nm and $r_{33} = 7.7 \pm 1.1$ pm/V at 1064 nm [123, 124]); in the telecom wavelength range around 1.55 µm we expect a value r_{33} of about 7 pm/V, since resonance effects are negligible already at 1064 nm, and an electro-optic figure-of-merit $n_3^3 \cdot r_{33} = 30$ pm/V.

4.3 Sample preparation

The electro-optic crystalline microresonators of COANP were fabricated using the recently developed melt-based channel growth technique [121] with a few modifications to improve the quality of the fabricated channels and electrodes. In a first step the waveguide structure and electrode pattern was fabricated on standard inorganic substrate materials. Subsequently, perpendicular to the substrate wafer plane the volume of the waveguide structure was delimited by anodic bonding of a cover. Finally, the organic material was melted at the edge of the cover, flowed through the obtained channels into the waveguide structure and crystallized there.

4.3.1 Microring resonator fabrication

As a substrate and cover material a commercial standard borosilicate glass suitable for anodic bonding was used. The borosilicate wafers had a thickness of 200 ± 25 µm and an average roughness of the surface of less than 1.5 nm. To manufacture the structured and electroded substrate wafer three photolithographic steps were required. In a first step 50 nm chromium was deposited and structured using electron-gun metal deposition and standard lift-off photolithography to form a floating electrode in the center of the ring and a split-ring-shaped electrode surrounding the microresonator (see Fig. 4.1a). The floating and the split-ring-shaped electrode was separated from the corresponding waveguide boundary by 5 µm. The two parts of this split-ring-shaped electrode were contacted via two separate 0.5 mm wide electrode structures, which were also patterned in this first fabrication step and which run parallel to the straight port waveguide of the final device. In the second lithographic step the microring resonator waveguide configuration was structured into the commercial photoresist AZ6632 and then transferred to the substrate wafer by subsequent reactive ion etching (RIE). A deep UV photolithographic system was used to achieve the close proximity of the port and ring waveguide in order to allow a reasonable mode coupling between the two waveguides. In the third lithographic step 50 nm silicon was deposited in regions where a permanent bond to the cover wafer had to be established via anodic bonding (see Fig. 4.1b). In order to enhance the adhesion of the silicon to the substrate wafer, a 40 nm thick layer of chromium was deposited prior to silicon during the same deposition cycle. As a slight variation of the described process roughly 180 nm spin-on-glass was spun onto the substrate after the first chromium deposition step to avoid the immediate contact between the electrodes and the organic material. Even though no rigorous analysis was made, we did not experience an essential difference in the crystal quality obtained with or without the spin-on-glass. Finally a cover borosilicate glass was anodically bonded to the structured substrate wafer as depicted in Fig. 4.1c. The parameters needed to bond the two wafers permanently together have been reported earlier [121]. The chromium and silicon deposited in the third lithographic step determined the bonding regions and at the same time they acted also as a spacer in regions where no bonding was established. Through the thereby generated channels the melt of the organic material was able to flow into the patterned waveguides and to efficiently fill the microring resonator structure.

As an option we included a straight coupling section in the resonator design, leading to a so called racetrack resonator.

4.3.2 Crystal growth

To flow the melt of the organic materials into the fabricated waveguide structures a few milligrams of COANP powder were placed onto the substrate wafer immediately at the edge of the cover glass. From there the patterned structures in the substrate glass were accessible for the melt due to the appropriate sample design detailed above.

To grow thin crystalline films of COANP the sample was put into a home-made double-wall chamber, whose temperature was controlled by water passing between its walls. By melting the entire amount of the initial COANP powder at the cover glass edge, the material converts into a glass form upon cooling to room temperature, which is due to a large difference between melting and crystallization points [125]. Therefore precise temperature tuning was required to prevent the melting of all the compound powder. The leftover powder acted as seed crystals, generating all possible domain orientations at the edge of the cover glass. However different growth rates for COANP crystalline films along the crystallographic b- and c-axes have been reported [125]. The microring structure is located very far from the edge of the cover glass compared to the width of the channel, which was used to flow the melt of the organic material

4. Electro-optic single-crystalline organic microring resonators

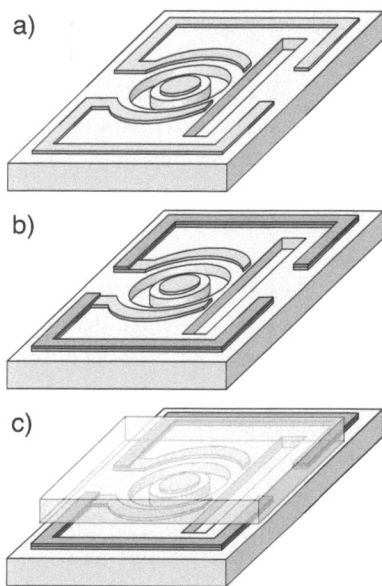

Figure 4.1: Processing steps for the fabrication of inorganic substrates for microring resonators. a) 50 nm chromium was deposited and structured to form the electrodes, which contoured the ring waveguide structure patterned by RIE. b) The deposited silicon defines the regions where the cover wafer is anodically bonded to the substrate and additionally acts as a spacer for the inner structure. c) A cover borosilicate glass shorter than the waveguide length was bonded to the structure to delimit the volume in vertical direction. A microscope image of the final device with the grown crystal in the waveguide channel is shown in Fig. 4.2.

into the structured waveguides (in Fig. 4.2 this channel contains the 140 nm thick nanosheet and has a width of about 380 μm). Therefore we most often observed only one crystalline domain of the nanosheet in the ring region with the fast growth direction c parallel to the long channel dimension (i.e. parallel to the port waveguide). Since the orientation of the crystalline domain of the nanosheet determines the corresponding crystal orientation in the waveguide channels, also the grown COANP crystals in the waveguides have their polar axis parallel to the port waveguide of the ring. Thus, this crystal orientation indicated in Fig. 4.2 accounts for the split-ring-shaped electrode configuration to apply the electric field mostly parallel to the polar axis. Since COANP crystallizes with its a axis perpendicular to the largest thin film crystal surface and exhibits point group symmetry $mm2$, light polarized parallel to the bc-plane is an eigenpolarization for any propagation direction parallel to the bc-plane.

A top view transmission microscope image of a COANP crystal grown from the melt in a racetrack resonator channel waveguide structured in a borosilicate substrate is depicted in Fig. 4.2. Figure 4.3 shows a cross-sectional view scanning electron micrograph of a typical

4.4. Transmission spectrum, resonance tuning and modulation

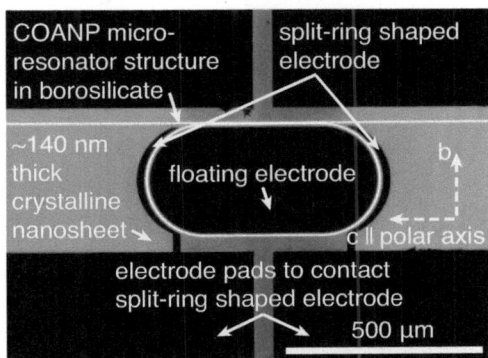

Figure 4.2: Top view transmission microscope image of a COANP racetrack resonator structure placed between crossed polarizers. The microresonator was accessible for the melt of COANP through a roughly 140 nm thick channel defined by the silicon/chromium spacer. The orientation of the crystal polar axis in the ring structure is determined by the orientation of the corresponding single-crystalline domain of the 140 nm thick nanosheet and is parallel to the port waveguide as indicated with dashed arrows. The corresponding cross-sectional view of a coupling region is shown in Fig. 4.3.

coupling region of such a racetrack resonator. Although we used deep UV lithography for the waveguide layout structuring, only a moderate proximity of the laterally coupled port and ring waveguides was reproducibly achievable. On account of this considerable gap in the coupling region of roughly 0.5 µm (see Fig. 4.3) a straight coupling section of racetrack resonators was used to increase the mode coupling between the waveguides. Figure 4.3 reveals also, that the surface roughness induced by the etch process is significantly reduced compared to our previous results on phase modulators [121], most likely due to a more optimal etch gas mixture.

To prepare for optical characterization, the device was cleaved from the fabricated sample perpendicular to the waveguide using a diamond scribe in order to couple light into the waveguide. In our previous work [121] we have shown that this cleaving technique produces vertical and smooth waveguide end-faces suitable for end-fire coupling and therefore the cleaved end-facets were left unpolished.

4.4 Transmission spectrum, resonance tuning and modulation

The electro-optic microring resonators with cross-sectional waveguide dimensions of about $w \times h = 3 \times 1.5$ µm^2 were characterized by transmission and electro-optic resonance tuning experiments. Polarized light was end-coupled into the waveguides by the conventional end-fire coupling technique using a 100× microscope objective. After propagating through the device, the out-coupled mode was projected either onto a CCD camera or a photodetector by means of a

4. Electro-optic single-crystalline organic microring resonators

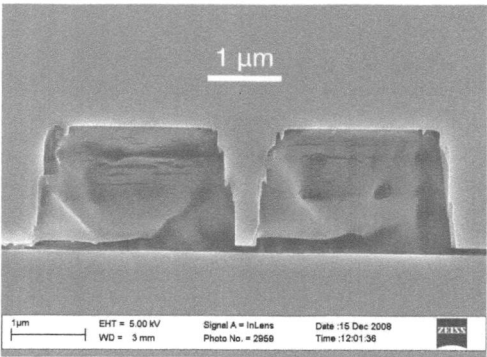

Figure 4.3: Cross-sectional view scanning electron micrograph of the coupling region of a racetrack resonator. A trapezoidal-shaped borosilicate gap with an average width of roughly 0.5 µm separates the two crystalline COANP waveguides. The imperfections seen in the end-facet of the waveguides are most likely caused by the electron-beam irradiation.

microscope objective with a 40× magnification. The transmission spectra of the microresonators were measured by scanning the wavelength of the input light from the tunable laser diode Santec TSL-220 and measuring the output power of the microring resonator. A LabView program was used for the simultaneous control of the tunable laser source and the corresponding recording of the power from the photodetector.

To analyse the measured spectrum we used the simple model describing an ordinary microring resonator filter with a single side-coupled waveguide acting as both input and output for the case of lossless coupling [126]. For a microring resonator of radius R, the round-trip phase delay is $\phi = \frac{2\pi}{\lambda} nL$, where $L = 2\pi R$, λ is the light wavelength and n is the effective mode index. After one round-trip the electric field in the resonator is reduced by $\xi = e^{-\alpha L/2}$, where α are the propagation losses. The coupling of the field between the waveguide and the ring can be described by coefficients τ and κ, where τ is the amplitude transmission coefficient across the coupling region, while κ describes the coupling into and out of the ring. In the assumed case of lossless coupling, the relation $\tau^2 + \kappa^2 = 1$ holds. Using these definitions, the transmission of the port waveguide can be expressed by

$$T = 1 - \frac{(1-\xi^2)(1-\tau^2)}{(1-\xi\tau)^2 + 4\xi\tau \sin^2(\phi/2)} \quad (4.1)$$

Note that the on-resonance transmission ($\phi = 2\pi \cdot m$, $m \in \mathbb{Z}$) drops to zero for the situation $\tau = \xi$. In this case, the internal losses are optimally compensated by coupling and the resonator is said to be critically coupled. For $\tau > \xi$, the resonator is said to be under coupled, and for $\tau < \xi$ the resonator is said to be over coupled. The finesse $\mathcal{F}$ of the resonator is a quantity commonly used to describe the sharpness of the resonance lines and is given by $\mathcal{F} = \pi\sqrt{\xi\tau}/(1-\xi\tau)$.

4.4. Transmission spectrum, resonance tuning and modulation

The measured TE spectrum of a racetrack resonator made of COANP with the straight section of the racetrack $S_{RT} = 150$ µm and the radius of the semi-circle $R = 150$ µm is shown in Fig. 4.4. It exhibits almost perfectly symmetric high extinction ratio resonance peaks of about 10 dB. The analysis of the presented measurement shows ring losses $\alpha = 12$ dB/cm ($\xi = 0.84$), a finesse $\mathcal{F} = 6.2$, a power coupling constant κ^2 between the port and resonator waveguide of 0.49 ($\tau = 0.71$) and a transmission's maximum slope $|dT/d\lambda|_{max}$ of about 6 nm^{-1}. According to the definitions above the racetrack resonator was over-coupled with a finite waveguide transmission of $T_{min} = 0.1$; for the given losses a power coupling constant $\kappa^2 = 0.29$ is required for critical coupling. Based on the absorption data reported for COANP [28], we estimated that the material absorption contributes about 4.3 dB/cm to the total ring losses, where we have assumed an isotropic absorption coefficient of 1 cm^{-1} for light with a polarization parallel to the bc-plane.

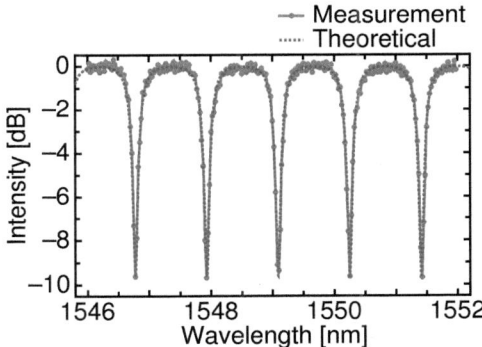

Figure 4.4: Transmission spectrum of a racetrack resonator with a straight section $S_{RT} = 150$ µm and a semi-circle radius $R = 150$ µm. The measured normalized transmitted light at the through port for TE modes using a tunable source in the $\lambda = 1.546 - 1.552$ µm region is shown. The free spectral range is 1.17 nm nm and the finesse 6.2. The modulation depth is approximately 10 dB.

For comparison, a circular microring with the same radius $R = 150$ µm featured extinction ratio resonance peaks of about 8 dB (see Fig. 4.5), a power coupling constant $\kappa^2 = 0.087$, a finesse $\mathcal{F} = 20$ and ring losses $\alpha = 10.3$ dB/cm ($\xi = 0.89$), of which approximately 4.3 dB/cm are again due to material absorption. In contrast to the racetrack resonator the microring was under-coupled and critical coupling for the given losses can be achieved with $\kappa^2 = 0.2$. For clarity, the gap in the coupling region was however different for the racetrack and microring resonator.

The electro-optic properties of COANP microrings have been investigated by shifting the transmission spectrum applying a static electric field to the device electrodes. The resonance of a TE mode displayed in Fig. 4.6 shows a $\Delta\lambda = 110$ pm-shift in response to an applied dc-voltage of 100 V, where the straight section of the racetrack was $S_{RT} = 300$ µm and the radius of the

4. Electro-optic single-crystalline organic microring resonators

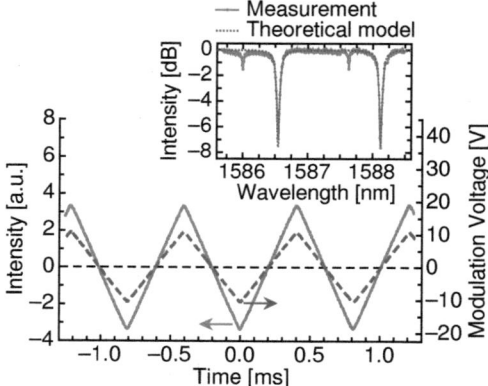

Figure 4.5: Electro-optic modulation seen at a repetition rate of 1.24 kHz by tuning the wavelength to the point of maximum slope of the through port TE response (shown in the inset) of a circular microring ($R = 150$ µm). The applied triangular modulation voltage to the electrodes is depicted as dashed curve and the modulated output intensity as solid line.

semi-circle was $R = 150$ µm. This wavelength shift corresponds to a frequency tunability of 0.13 GHz/V (1.1 pm/V), which is comparable to what has been reported for ion-sliced LiNbO$_3$ microring resonators [34].

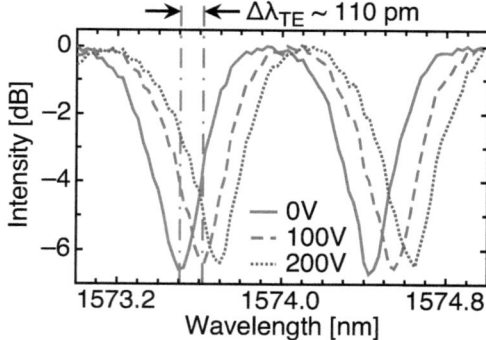

Figure 4.6: Resonance spectrum of a TE mode at a wavelength around 1.57 µm (solid line) of a racetrack resonator; the dashed and dotted line are the corresponding electro-optically shifted curves by applying a voltage $V = 100$ V and $V = 200$ V to the device electrodes respectively. The shift indicated by vertical dashed-dotted lines corresponds to a tunability of about 0.13 GHz/V (1.1 pm/V).

We also demonstrated the electro-optic modulation for a fixed wavelength in the steep flank of a microring resonance by applying a triangular modulation voltage at a repetition rate of 1.24 kHz to the two parts of the split-ring-shaped electrode. The applied modulation voltage and the resulting light intensity modulation at the output port were monitored with an oscilloscope. Especially the circular microring with its very high finesse of about 20 was very well suited for electro-optic modulation experiments. With the induced transmission modulation and the slope of the spectrum at the modulation wavelength known, the ac-modulation experiments can be compared to the dc-induced wavelength shifts of the spectrum. The tunability obtained from the electro-optic modulation experiments was generally lower by a factor of about 1.4 compared to resonance tuning. A possible explanation of this observation can be given in terms of an increase of the dielectric constant of borosilicate cladding due to a dispersion effect towards very low frequencies. This may lead to a higher effective electric field, which interacts with the organic crystal waveguide for dc-measurements compared to ac-modulation and accounts for the larger dc-induced tunability.

4.5 Discussion and outlook

In the following, we estimate the expected electro-optic tunability of the single crystalline COANP microring resonators with the crystal orientation indicated in Fig. 4.2 and an electric field applied to the split-ring shaped electrodes. The electric-field-induced phase change $\Delta\phi$ of the phase $\phi = \phi_0 + \Delta\phi$ in Eq. (4.1) accumulated during one round-trip of the optical mode can be calculated as

$$\Delta\phi = \int_{-\pi/2}^{\pi/2} \frac{4\pi R}{\lambda} \Delta n(\theta) \, d\theta \tag{4.2}$$

$$= -\frac{4\pi R E_0}{\lambda} \int_0^{\pi/2} \frac{(r_{23} + 2r_{42})\sin^2\theta + r_{33}\cos^2\theta}{\left(\frac{\sin^2\theta}{n_2^2} + \frac{\cos^2\theta}{n_3^2}\right)^{3/2}} \cos\theta \, d\theta \tag{4.3}$$

where we have considered the deformation of the optical indicatrix due to a radially applied electric field E_0 in the bc-plane of COANP, where ϕ_0 is the unperturbed phase, θ is the angular coordinate and $r_{mk} = r_{ijk}$ ($m = 1, \ldots, 6$) are the electro-optic tensor elements given in the usual contracted notation. To determine the electric field $E_0 = V/d_{\text{eff}}$ in the crystalline COANP waveguide, where V is the applied voltage, we have considered the effective thickness d_{eff} between the electrodes given by

$$d_{\text{eff}} = 2 \cdot \varepsilon_{\text{COANP}} \cdot \left(\frac{5\,\mu\text{m}}{\varepsilon_{\text{BS}}} + \frac{3\,\mu\text{m}}{\varepsilon_{\text{COANP}}} + \frac{5\,\mu\text{m}}{\varepsilon_{\text{BS}}}\right) \tag{4.4}$$

which takes into account the difference between the dielectric constant of COANP ($\varepsilon_{\text{COANP}} = 2.7$ at 1.59 kHz [124]) and of borosilicate ($\varepsilon_{\text{BS}} = 4.6$ at $25°C$ and 1 MHz [127]). The factor 2 in Eq. (4.4) is because the applied voltage drops twice over the distance between the split-ring-shaped electrode and the floating electrode. By using the values $r_{33} = 7$ pm/V, $r_{42} \approx r_{23} = 5.4$ pm/V, $n_2 = 1.7$, $n_3 = 1.6$ [123, 124] and an average refractive index

$$\bar{n} = \frac{1}{2\pi} \int_0^{2\pi} \left(\frac{\sin^2\theta}{n_2^2} + \frac{\cos^2\theta}{n_3^2}\right)^{-1/2} d\theta \approx 1.65 \tag{4.5}$$

involved in the unperturbed phase $\phi_0 = \frac{2\pi}{\lambda}\bar{n}L$, where L is the round-trip propagation length, we get with the help of Eq. (4.3) and (4.4) a tunability of 0.092 GHz/V (0.76 pm/V) at a wavelength of about 1.57 µm. This theoretically expected value agrees very well with the measured tunability in the ac-modulation experiments, which was about 0.095 GHz/V (0.79 pm/V). As detailed above, the dispersion of the dielectric constant of borosilicate is probably the reason for the observed higher dc-tunability of 0.13 GHz/V (1.1 pm/V), due to the enhancement of the electric field in the COANP waveguide, i.e. decrease of d_{eff} according to Eq. (4.4) at larger ε_{BS}.

Note, that in the calculation we have considered an interaction length of the applied electric field and the organic crystal waveguide, which corresponds to the total circumference of the microring. Numerical calculations have shown that the tunability obtained for the actually employed electrode layout is only different by a few percent from the value we reported above. On the other hand the conceptually interesting microring design with vertical electrodes and an appropriate COANP crystal orientation with the polar axis perpendicular to the film surface would lead to a tunability comparable to the present crystal and electrode configuration.

Further efforts are needed to fully unleash the potential of microring resonators based on organic crystalline materials. A first very important improvement relates to the tunability. To achieve more efficient microring resonators a higher electro-optic tunability is required, in order to achieve a larger electro-optically induced resonance shift per unit voltage. The most obvious approach to increase the demonstrated tunability of 0.13 GHz/V (1.1 pm/V) by a factor of 15 is by using one of the presently best melt-processable organic materials with a correspondingly higher electro-optic figure-of-merit such as OH1 (2-{3-(4-hydroxystyryl)-5,5-dimethylcyclohex-2-enylidene}malononitrile) ($n_3^3 \cdot r_{33} = 450$ pm/V at 1.55 µm) [120].

Another interesting improvement is the possibility to realize ring resonators with smaller radii. For example high-contrast semiconductors in particular silicon-on-insulator are ideally suited for ring radii even below 2 µm [128]. On the other hand, polymer based microrings are limited to larger ring radii due to their low refractive index contrast compared to crystalline organic microrings with their generally larger refractive index values. Evidentially the refractive indices of COANP are comparable to those of standard electro-optic polymers, nevertheless the refractive index contrast between borosilicate glass and the presently best nonlinear optical organic crystals such as OH1 [120] or DAPSH (trans-4'-(dimethylamino)-N-phenyl-4-stilbazolium hexafluorophosphate) [116] reaching refractive-index values of up to 2.3 at 1.55 µm, allows for ring radii below 10 µm. For smaller rings the optical mode in the microring resonator is, however, closer to the outer edge of the ring and is therefore more sensitive to sidewall roughnesses. Furthermore, smaller rings imply also shorter coupling regions between port waveguide and ring. Therefore more advanced lithography of glass substrates should allow to precisely adjust the gap size as well as to reduce the scattering losses for the case of reduced ring dimensions.

Alternatively, organic single-crystalline materials have a great potential to be used in combination with silicon-on-insulator waveguides to electro-optically tune the guided modes. This alternative modulation scheme potentially allows to overcome the presently fastest 30 GHz bandwidth modulation achieved with all-inorganic injection based silicon-on-insulator modulators [14], in order to meet todays requirements for ultra-high bandwidth performance using the ultra-fast electro-optic response in organic materials.

4.6 Conclusion

We have fabricated microring resonators with single-crystalline COANP as electro-optic active material. The realization of microring resonators depends crucially on the high-precision fabrication and structuring of waveguides. We have employed and optimized the recently developed technique, in which the electro-optic organic crystals are grown from the melt in electrode equipped waveguide channels patterned with conventional optical lithography, standard lift-off technique, RIE and terminated by an anodic wafer bonding step. The accuracy and reproducibility of the process is largely increased compared to direct structuring techniques, since the entire photolithographic processing is applied to high-quality inorganic substrates.

With this optimized fabrication technique, we could realize the first electro-optic single-crystalline microring resonator in an organic material. For the first demonstration, we have chosen conventional optical lithography for the sample fabrication due to its simplicity. COANP was chosen as electro-optic active material, since capillary methods allow the growth of high-quality thin film crystals from the melt. The realized microring and racetrack resonators featured a high extinction ratio of 10 dB, moderate dimensions with a ring radius of $R = 150$ µm and a finesse of up to 20. An approximate tunability of 0.13 GHz/V (1.1 pm/V) was found by electro-optically shifting the resonance curve by applying a dc-voltage to the device electrodes and electro-optic modulation was observed for a fixed wavelength in the steep flank of a microring resonance.

The first electro-optic microresonators based on the crystalline organic material COANP fabricated in this work are a very concrete proof for the applicability of submicron scale feature patterning of crystalline organic materials.

4. Electro-optic single-crystalline organic microring resonators

CHAPTER 5

Conclusions and outlook

5.1 Conclusions

In this thesis, a new melt based channel growth technique has been developed and successfully applied in order to realize organic electro-optic single-crystalline integrated waveguide modulators, microresonators and nanowires.

For the development of very large-scale integrated photonic devices, high-quality noncentrosymmetric organic thin films are required, since such quasi two-dimensional films are highly compatible with planar waveguide configurations. Because of this, polymers are most commonly employed, since they are easy to process in thin films and offer a relatively high freedom to exploit different device production techniques. Nevertheless they often show thermal and photochemical instabilities and their nonlinearities are generally lower than in the crystalline counterparts. On the other hand, organic single crystals have superior thermal and photochemical stability and do not require high-field poling, but are difficult to process, especially in thin films. Thus, different techniques for the fabrication of high-quality organic single-crystalline thin films have been developed in the past. To obtain two-dimensional organic crystals, especially the growth by capillary methods between two plates has been repeatedly investigated. However, previous attempts to grow crystalline thin films from the melt between planar glass plates, encountered problems with the removal of one of the cladding plates after crystal growth, which often led to mechanical damage of the thin crystalline layers. In order to allow for post-growth cleaving of the samples, in this work the plates were permanently sealed by a wafer bonding technique, otherwise commonly used in microsystem technology.

To fabricate integrated photonic devices a lateral light confinement in the organic thin films has to be obtained. The difficulties to fabricate organic single-crystalline waveguides often arise due to the incompatibility of the solvents used for conventional optical lithography with the organic thin films. In the developed channel growth technique, the waveguide channels were structured into the inorganic substrate plate and equipped with electrodes prior to bonding,

5. Conclusions and outlook

which enabled the accurate control of the geometrical channel dimensions for the subsequent crystallization of the organic material therein. Consequently, one of the main advantages of the investigated structuring technique, compared to standard channel waveguide manufacturing, is that neither photolithography nor chemical or mechanical etching has to be directly applied to the organic crystals. Therefore, the side wall roughness of the finally obtained organic waveguides are generally much smoother compared to the one produced by standard processing.

A drawback of the developed method is that the growth from the melt is encountered by the characteristic difficulty of the formation of cracks within the layer, which arises from a difference in the thermal expansion coefficients for the organic layer and the inorganic material where the waveguide channels have been patterned. In order to obtain a small thermally induced strain in the grown crystals and consequently high quality thin films, COANP with its relatively low melting point has been used to fabricate the microring resonators. Another problem with the channel growth technique is to obtain the proper crystallographic growth direction with respect to the use of large electro-optic tensor elements. Previous experimental approaches for crystal cored organic fibers have shown that it is nearly impossible to influence the natural growth direction with respect to the fiber axis. Unfortunately, most of the potentially interesting electro-optic materials crystallize with their polar axis almost parallel to the straight waveguide axis and are therefore ruled out for their use in electro-optic waveguide phase and Mach-Zehnder modulators. As one of a few known exceptions, DAT2 crystallizes with its polar axis perpendicular to the waveguide channels, which allowed to realize phase and Mach-Zehnder modulators. On the other hand, different growth rates for crystalline COANP films along the crystallographic axes combined with the circular nature of the ring and an appropriate electrode design, allowed to exploit the largest electro-optic coefficient for modulation of the active microring resonators.

After the development of a suitable fabrication method for the realization of single-crystalline waveguide structures, the successful demonstrations of electro-optic Mach-Zehnder modulators and microring resonators marks an important step towards the utilization of organic crystals in integrated devices for telecommunication applications. Nevertheless, at present the half-wave voltage × length product of the electro-optic Mach-Zehnder modulators is with 60 Vcm still relatively high. Also the tunability obtained with the first active single-crystalline organic microring resonators of about 0.13 GHz/V would have to be improved, in order to become technologically relevant. Since the respective materials used in the devices had a comparably low electro-optic coefficient, there is, however, a very obvious approach to improve the relevant device parameters by one order of magnitude, by using one of the presently best melt processable electro-optic materials, such as OH1.

Beside integrated modulator device fabrication also more basic research on the development and the characterization of new organic crystals has been made. Among the studied materials, DAPSH has shown a second order susceptibility of 580 pm/V measured at a wavelength of

1.9 µm, which is presently the largest non-resonant nonlinear optical coefficient reported in the literature.

5.2 Outlook

In this thesis, phase and amplitude electro-optic modulators, as well as first microresonator filters and modulators based on organic electro-optic crystals were demonstrated. For this a new melt-growth technique to obtain organic crystalline micro- and nanostructures inside large-area devices was developed. The new technique is very versatile and can therefore find use also in other areas, where functional organic materials need high integration, e.g. for light emitting or conducting on-chip applications.

In the field of very large-scale integrated optics, silicon has emerged in the last few years as the most promising material platform for further miniaturization. Since silicon does not exhibit intrinsic electro-optic activity, a very effective approach is offered by hybrid integration of nonlinear optical organic materials and silicon waveguides. Organic electro-optic materials introduced into silicon waveguide outlines could be advantageous for electro-optic modulators, due to their potential for low switching voltages and for ultra-fast tuning. Therefore, organic crystalline thin films and nanosheets obtainable with the developed melt-based channel growth technique are extremely interesting for their integration in a wide range of silicon photonics structures. In the most simple approach an organic single crystalline thin film could be used as an electro-optic active cladding to silicon nanowire waveguides in order to electro-optically tune the guided modes. In more sophisticated device configurations, organic crystals could be infiltrated into photonic bandgap structures or slotted silicon waveguide configurations; especially the latter has attracted a lot of interest in the recent past. In general, it is relatively difficult to fill the very narrow slots using conventional organic material processing techniques, but in this thesis it has been demonstrated that the melt-based channel growth technique can also be used to fabricate single-crystalline nanostructures inside large-area devices with crystal thicknesses below 30 nm and lengths of above 7 mm. Therefore, it is very appealing to grow crystalline nanosheets in slotted silicon waveguide structures, also because the fabrication techniques used in silicon photonics can be very similar to those used to manufacture the substrates in this thesis. For example, uninfiltrated horizontal slot waveguides based on bonded silicon-on-insulator wafers have been very recently reported [129], where an amazingly similar production technique has been used to fabricate the channels, as we have used to demonstrate the growth of the nanosheets. Such prospective configurations are interesting options to combine the matured passive silicon-on-insulator technology with the ultra-fast electro-optic activity and highly stable chromophore orientation of organic crystalline materials. These silicon/organic hybrid devices hold great promise for denser integration, higher switching speeds and lower operating voltages than the inorganic state-of-the-art material, $LiNbO_3$.

5. Conclusions and outlook

APPENDIX A

Appendix to chapter 2

A.1 Maker fringe measurements

The nonlinear optical tensor elements were measured by the standard Maker fringe technique where the corresponding theoretical model to analyze the data was based on the method developed by Herman and Hayden, in which both the absorption and birefringence of the crystal as well as reflections of the second harmonic wave were considered [130]. The fundamental beam at 1.907 µm was generated by focusing a Q-switched Nd:YAG laser at 1.064 µm with a pulse energy of about 150 mJ and a repetition rate of 10 Hz ($\tau = 25$ ms) into a high pressure Raman cell filled with H_2. Thereafter the beam was focused with a lens of 65 mm focal length on the crystal, which was mounted on a rotation stage, whose rotation axis intersects the beam line. The generated second harmonic light was measured with a photo multiplier and integrated with a boxcar integrator. The acquisition of the SHG intensities and the rotation of the crystal were computer controlled. All measurements were referenced to quartz with $d_{111} = 0.277$ pm/V at 1.907 µm [66].

The nonlinear polarization $\mathbf{P}^{2\omega}$ used for SHG of an incident field $\mathbf{E}^{\omega} = (E_1, E_2, E_3)$ at the fundamental frequency ω in a material of point group m is given by

$$P_1^{2\omega} = \varepsilon_0(d_{111}E_1^2 + d_{122}E_2^2 + d_{133}E_3^2 + 2d_{113}E_1E_3) \tag{A1}$$

$$P_2^{2\omega} = \varepsilon_0(2d_{223}E_2E_3 + 2d_{212}E_1E_2) \tag{A2}$$

$$P_3^{2\omega} = \varepsilon_0(d_{311}E_1^2 + d_{322}E_2^2 + d_{333}E_3^2 + 2d_{313}E_1E_3), \tag{A3}$$

where ε_0 is the electric permittivity of free space and d_{ijk} the components of the nonlinear optical tensor

$$d = \frac{1}{2}\chi^{(2)}(-2\omega, \omega, \omega) , \tag{A4}$$

where $\chi^{(2)}$ is the second-order nonlinear optical susceptibility.

A. Appendix to chapter 2

Note that for birefringent materials the electric field vectors are not perpendicular to the wave vectors **k**. Therefore, for the evaluation of the data the walk-off angles between the Poynting vectors and the wave vectors have to be taken into account.

For general directions of the wave vector and the polarizations in the crystal, the projection of the induced polarization at frequency 2ω along the direction of the electric field of the generated wave at 2ω can be written as

$$|\mathbf{P}^{2\omega}| = 2\varepsilon_0 d_{\text{eff}} |\mathbf{E}^{\omega}||\mathbf{E}^{\omega}| \,, \tag{A5}$$

with

$$d_{\text{eff}} = \sum_{ijk} d_{ijk}^{(2\omega,\omega,\omega)} \cos(\alpha_i^{2\omega}) \cos(\alpha_j^{\omega}) \cos(\alpha_k^{\omega}) \,, \tag{A6}$$

where α_i^{ω} is the angle between the electric field vector at frequency ω and the axis i of the Cartesian coordinate system in which d_{ijk} is given [131].

A.2 Refractive index measurement

The refractive indices of DAPSH were measured with an interferometric method [28, 86], which is based on changing one optical path length in the Michelson interferometer by rotating the samples.

The polarization state of the input light was either parallel (s-polarization) or perpendicular (p-polarization) to the axis of rotation. By rotating the crystal in the beam line, the optical path length changed. This change in the optical path length depends on the index of refraction, leading to a change in the interference pattern of the two beams in the Michelson interferometer. Thus, the refractive indices were determined by recording the interference fringes as a function of the rotation angle [28, 86]. A photodiode was used to record the light intensity at the exit of the Michelson interferometer. Both the rotation of the sample and the simultaneous measurement of the light intensity were computer controlled.

Four unknown parameters (n_1, n_2, n_3, φ) had to be determined, where φ is the inclination angle of the dielectric system with respect to the normal to the bc plane, analogous to ψ in Fig. 2.1b), and n_1, n_2 and n_3 are the corresponding refractive indices. We consider the general case, where φ might be different from ψ.

In a first measurement the crystal was mounted such that the dielectric axis x_2 and the rotation axis as well as the polarization direction were all parallel. In a second measurement the polarization was rotated by $90°$. For these two configurations, the rotation axis was parallel to the dielectric axis x_2, and it was straightforward to find a relation between the number of interference fringes m and the rotation angle ϕ, since both s- and p-polarized light are eigenpolarizations in the crystal for every rotation angle. In a third measurement the polarization was again parallel to the rotation axis and the crystal was rotated by $90°$ in the bc plane such that the crystallographic axis c was parallel to the rotation axis. In this configuration, waves with different polarization states are produced, due to the anisotropy of the crystal. In

A.3. Refractive indices and coherence length

order to calculate the number of interference fringes (or in other words the phase difference with respect to the position of normal incidence) in this more general case, an approach analogous to the 4×4 matrix formalism of Berreman [132] was used.

The refractive index n_2 was obtained from the first measurement, while the three remaining parameters (n_1, n_3 and φ) were determined by simultaneously fitting the number of interference fringes from the second and third measurement as a function of the rotation angle ϕ. This procedure was repeated for several wavelengths. Figure A1 shows a data set with a least square fit to the number of interference fringes, where the inclination angle φ was kept constant at 55.8° and the measurement was taken at a wavelength of 1.064 µm. Several measurements showed that the inclination angle φ is between 53.8° and 57.8° with respect to the surface normal of the crystal, therefore we decided to set $\varphi = 55.8°$ according to the orientation of the molecules shown in Fig. 2.1b), i.e. having $\varphi = \psi$. Therefore, the dielectric system corresponds exactly to the Cartesian system defined in Fig. 2.1b), 2.1c) and 2.2b).

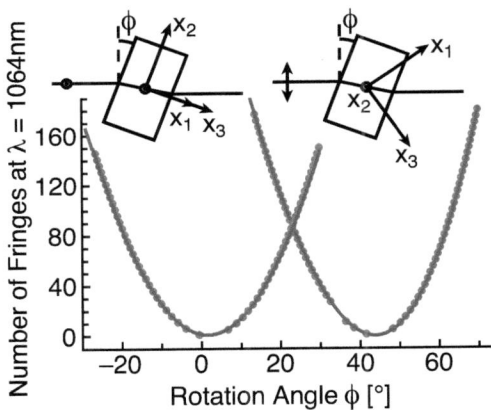

Figure A1: Number of interference fringes as a function of the external rotation angle (0° indicates normal incidence) for two experimental configurations as illustrated above the corresponding parabolas. For reasons of clarity only every fifth data point is shown. For the theoretical curve (solid line) the refractive indices n_1 and n_3 were varied, whereas the inclination angle φ was fixed at 55.8°. The shift to positive angles of the parabola measured with p-polarized light is due to a non-zero inclination angle φ between the bc surface and the dielectric system.

A.3 Refractive indices and coherence length

The positions of the minima of the Maker fringes as well as the incidence angle, at which the phase matched second harmonic signal are observed, provide precise information about the refractive indices and their dispersion. It is interesting to compare these results to the ones obtained from the interferometric refractive index measurements. From Eq. (2.1) and Table 2.1

A. Appendix to chapter 2

we get $n_2^\omega = 1.58 \pm 0.07$ and $n_1^{2\omega} = 2.46 \pm 0.05$, where n^ω and $n^{2\omega}$ denote the refractive indices at the fundamental and second harmonic frequency respectively. Phase matching was observed at an angle of $(41 \pm 1)°$, which gives $n_3^{2\omega} = 1.432 \pm 0.009$ in good agreement with the result shown in Table 2.1 ($n_3 = 1.45 \pm 0.02$). From the positions of the minima of the Maker fringes the refractive indices n_1^ω and n_3^ω can be determined by a least square analysis. For a minima the following condition holds:

$$\frac{\Delta k(\phi) \cdot d}{2} = N(\phi)\pi , \tag{A7}$$

where ϕ is the external angle of incidence, d is the thickness of the crystal, $N(\phi)$ is an integer number and $\Delta k(\phi)$ is given by

$$\Delta k(\phi) = \frac{2\omega}{c} \left| n^\omega(\phi'_\omega) \cos \phi'_\omega - n^{2\omega}(\phi'_{2\omega}) \cos \phi'_{2\omega} \right| , \tag{A8}$$

where $\phi'_{\omega,2\omega}$ denote the internal angles for the fundamental and second harmonic light. If the refractive indices $n_1^{2\omega}$ and $n_3^{2\omega}$ determined above are used in Eq. (A7) and (A8), we get $n_1^\omega = 2.22 \pm 0.07$ and $n_3^\omega = 1.429 \pm 0.006$. For the computed refractive indices the number of minima of the measured and calculated curve are in very good agreement, as shown in Fig. 2.6b). The refractive indices are depicted as diamonds in Fig. 2.3 and were also used for the evaluation of the nonlinear optical susceptibilities from the Maker fringe experiments.

APPENDIX B

Appendix to chapter 3

B.1 Refractive index measurement

The determination of the refractive indices in DAT2 is not trivial because of an inclined indicatrix with respect to the crystal faces, therefore three independent measurement methods were carried out. The first method was performed on solution grown thin films due to their excellent crystal quality and large area of up to 1×10 mm^2 [103]. For these crystalline thin films it is known that the largest crystal face is parallel to the ab-crystallographic plane (c-face) [103]. The directions of the crystallographic a and b axes can be for example determined by polarized second-harmonic generation measurements, where it was found that the a axis is parallel to the long axis of the rectangular crystals grown from the solution. In the point group symmetry 2, the dielectric x_2 axis is set parallel to the crystallographic b axis. Therefore light polarized parallel or perpendicular to the long axis of the crystal is an eigenpolarization for normal incidence to the as grown c-plate crystals.

A normal incidence transmission spectrum of such a crystalline thin film on a thick glass substrate (1 mm) was recorded with a Perkin-Elmer Lambda9 spectrometer. The spectrum features up to 33 interference fringes in the wavelength range from 600 to 2000 nm due to multiple coherent reflections in the film. A thin film with a thickness $d = 6.90$ µm was used for this measurement. The thickness was determined by a Tencor Alphastep P11 profilometer yielding an accuracy of ± 50 nm. The actual thickness value used for the evaluation of the data was obtained from a prism coupling experiment and lay within the error tolerance (see below). The wavelength values of the interference minima and maxima were extracted for the evaluation of the refractive index dispersion. The relation for a fringe minima or maxima is simply the expression for destructive or constructive interference after one roundtrip in the thin film: $2dn(\lambda)/\lambda = p$ or $2dn(\lambda)/\lambda = q + 1/2$ respectively, where p and q are integer numbers and denote the order of the minima or maxima. Figure B1 shows the extracted fringe maxima as a function of wavelength for the two accessible eigenpolarizations. According to the crystal

B. Appendix to chapter 3

orientation described in the main text, n_2 denotes the refractive index for light polarized parallel to the crystallographic b-axis (i.e. the dielectric x_2-axis) and n_{13} is the refractive index for light propagating normal to the ab-plane and polarized parallel to the crystallographic a-axis. The notation of n_{13} reflects the fact, that the main refractive indices n_1 and n_3 contribute to n_{13} due to a possible tilt of the indicatrix in the ac-plane with respect to the c-face. The dispersion of the refractive indices $n(\lambda)$ is described with a simple Sellmeier one-oscillator model given by

$$n^2(\lambda) = n_0^2 + \frac{q\lambda^2}{\lambda^2 - \lambda_0^2} = n_0^2 + \frac{E_d E_0}{E_0^2 - E^2} \;, \tag{B1}$$

where $\nu_0 = c/\lambda_0$ is the resonance frequency of the main oscillator, q is the oscillator strength, and n_0 is a constant depending on the contributions from all other oscillators. The expression on the right-hand side is the corresponding energy description, with the oscillator energy $E_0 = h\nu_0$ and the oscillator strength $E_d = q E_0$. The described analysis was performed for the two accessible eigenpolarizations parallel to the a and b axis, which yields the dispersion of the corresponding refractive indices n_{13} and n_2 respectively. The full and dashed lines in Fig. B2 are according to Eq. (B1) obtained with the parameters given in Table B1.

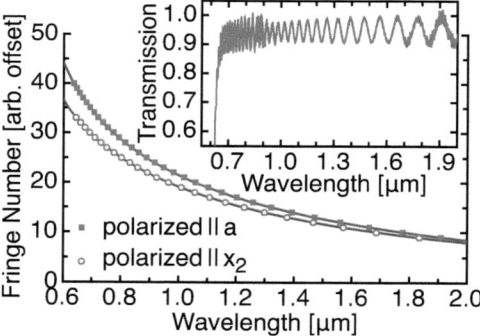

Figure B1: Numbered fringe maxima of normal incidence transmission spectrum measured on a DAT2 crystalline thin film. The light was polarized parallel to the dielectric x_2-axis (open dots) or perpendicular to it (i.e. parallel to the crystallographic a-axis) (full squares). The solid lines correspond to the Sellmeier model in Eq. (B1) with parameters of Table B1. The inset shows a typical transmission spectrum measured, from which the fringe maxima were extracted.

The refractive indices calculated from the transmission spectrum of a thin film were confirmed by a birefringence measurement using a tilting compensator B (Zeiss) with a polarizing microscope (Zeiss) and a white light source. The investigated birefringence is the difference between the refractive index n_{13} and the refractive index n_2: $\Delta n = n_{13} - n_2$. For a proper crystal orientation both eigenpolarizations could be excited and the generated phase difference was compensated by the tilting compensator. From the known optical retardation of the compensator, the birefringence could be calculated. The birefringence expected from the

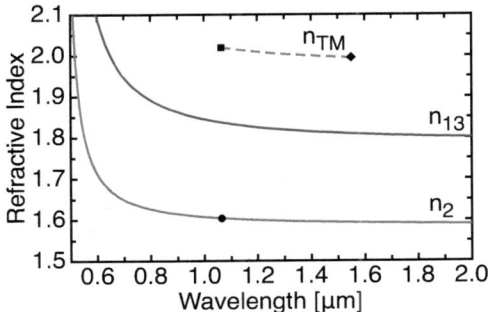

Figure B2: Dispersion of the refractive indices of DAT2 in terms of a one-oscillator model. The solid lines represent the refractive index experienced by light polarized parallel (n_2) and perpendicular (n_{13}) to the dielectric axis x_2 with normal incidence to the c-face of a DAT2 crystal as a function of the wavelength. The solid curves were obtained from a least square theoretical analysis of the Sellmeier model in Eq. (B1) to the data from a normal incidence transmission spectrum. The discrete data points at 1.064 µm represent the resulting refractive indices determined from a prism coupling experiment. To obtain the refractive index value experienced by a TM mode at 1.55 µm (filled diamond) the corresponding data point measured at 1.064 µm (filled square) was extrapolated assuming the same dispersion coefficients q and λ_0 (dashed curve) as for the refractive index n_{13}.

Table B1: Sellmeier parameters for the refractive index dispersion of Eq. (B1) which correspond best to the experimental data obtained from a normal incidence transmission spectrum or which were used to extrapolate the refractive index measured for a TM mode at 1.064 µm to a wavelength of 1.55 µm. The given values are the parameters for calculating the refractive indices, while the error indicated is the error range of the measurement.

	n_{13}	n_2	n_{TM}
q	0.6887 ± 0.09	0.2324 ± 0.06	0.6887 ± 0.11
n_0	1.586 ± 0.04	1.511 ± 0.03	1.794 ± 0.07
λ_0 [nm]	476.01 ± 15	480.83 ± 30	476.01 ± 20
E_0 [eV]	2.61 ± 0.08	2.58 ± 0.16	2.61 ± 0.11
E_d [eV]	1.798 ± 0.24	0.600 ± 0.16	1.798 ± 0.30

dispersion parameters given in Table B1 is depicted as dotted line in Fig. B3. The value determined from the compensator measurement is shown as dashed line. The two birefringence values are in close agreement immediately at the absorption edge of the crystal, since shorter wavelengths of the white light source are absorbed in the crystal.

The second method is based on a prism coupling measurement, for which solution grown thin films were used as well. Prism coupling is a non-destructive and powerful technique to investigate the optical properties of thin films, in particular, to measure the refractive index and

B. Appendix to chapter 3

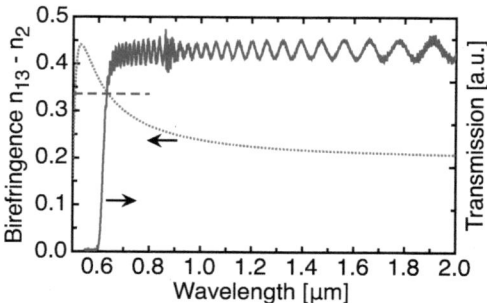

Figure B3: Birefringence between the two eigenpolarizations for incident normal to the ab-crystallographic plane. The dotted curve represents the birefringence calculated from the dispersion parameters in Table B1. The dashed line is the birefringence measured with a tilting compensator B and a polarizing microscope with a white light source. The two birefringence values are in good agreement close to the absorption edge in the transmission spectrum of DAT2 (solid line).

the thickness simultaneously. A setup consisting of a right-angle rutile (TiO_2) prism was used to couple a Nd:YAG-laser beam (1.064 µm) to the guided modes of a thin film crystal. The wave vector of the excited waveguide modes was in all experiments parallel to the crystallographic a-axis. A properly oriented film was pressed against a cathetus of the asymmetric prism, having an angle of $\delta = 50°$ between this cathetus and the hypotenuse. In the experiment, we measured the reflected intensity as a function of the angle of incidence. Under certain incidence angles, sharp reflectivity dips appear in the recorded spectrum, which correspond to the excitation of guided modes. The refractive indices were determined by using light of transverse electric (TE) and transverse magnetic (TM) polarization, from the angular position for the excitation of TE or TM guided modes. By Snell's law at the prism-film interface, the measured coupling angle φ_m with respect to normal incidence on the hypotenuse determines the effective mode index N_m by

$$N_m = n_p \cdot \sin\left(\delta - \arcsin\left(\frac{\sin(\varphi_m)}{n_p}\right)\right) , \qquad (B2)$$

where n_p is the prism refractive index and δ the mentioned angle of the prism. The refractive indices of the rutile prism used for the evaluation of the measurements were taken from Ref. [133]. The refractive index of the film has been deduced from the effective refractive indices N_m by numerically solving the conventional transcendental equation describing an asymmetric step-index planar waveguide.

The prism coupling measurement was performed with a thinner and a thicker crystal, with a thickness of 0.97 ± 0.05 µm and with 22.1 ± 0.1 µm respectively. Figure B4 shows the extracted coupling angles as a function of the mode number for the thicker crystal for TE and TM polarization. The refractive index values determined for the two crystals were in good agreement and have a smaller absolute error compared to those obtained from the normal

B.1. Refractive index measurement

incidence transmission spectrum. The thickness of the crystal was also a parameter in the simultaneous least square analysis of the normal incidence transmission spectrum and prism coupling angles measured for light polarized parallel to x_2 and it was found that the thickness in best agreement with the experimental data was within the error tolerance of the profilometer measurement. This is a good proof for the high reliability of the determined refractive indices. The refractive index value measured at a wavelength of 1.064 µm for TM polarization (n_{TM}) was extrapolated to larger wavelengths using the dispersion coefficients given in Table B1.

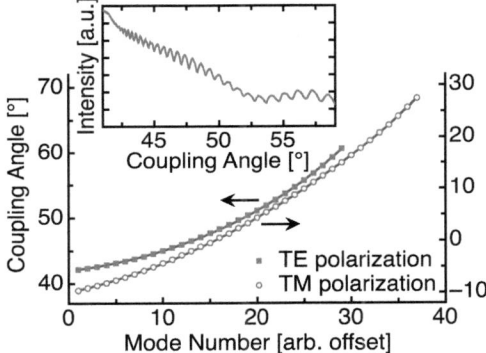

Figure B4: Coupling angle measured in a prism coupling experiment with a DAT2 crystalline thin film. The wave vector of the guided modes was along the crystallographic a-axis and the light was polarized parallel to the dielectric x_2 axes (TE-modes, full squares) or polarized perpendicular to the crystallographic ab-plane (TM-modes, open dots). The solid lines are the least square theoretical fits to a model based on Eq. (B2) and the standard mode equation of an asymmetric step-index planar waveguide. The inset shows a typical reflected intensity in a coupling experiment as a function of the incidence angle on the hypotenuse of the prism. Note that prism coupling for TM and TE polarization was performed with the crystal on glass substrate and without a substrate respectively.

For the third method, which is based on conoscopy, we used a bulk, ∼150 µm thick, crystal grown by slow evaporation from acetone solution. Conoscopy is an interferometric technique and it is among others commonly used for the detection of the orientation of optical axis. Conoscopic interference fringes are generated for anisotropic crystals viewed between two crossed polarizers with a converging light beam. Basically the two eigenpolarizations excited in the crystal interfere and produce the characteristic fringe pattern. In a plane normal to the optical axis direction of the crystal, the constant phase curves are concentric fringes (isochromes). The crystal was illuminated with a polarized Nd:YAG laser beam at a wavelength of 1.064 µm through a high numerical aperture lens which created the convergent light beam onto the crystal. A second high numerical aperture lens after the crystal collected as much light as possible. A crossed analyzer with respect to the laser light polarization was placed after the lens. The experiment revealed that the optical axes are in the ac-plane since concentric fringes were

B. Appendix to chapter 3

observed for a rotation of the crystal around its b-axis. The corresponding incidence angle with respect to the sample normal was $\phi' = 65 \pm 2°$ at a wavelength of 1.064 µm. Moreover, also the characteristic melatope for biaxial crystals was clearly visible in the center of the isochromes.

From these three independent experiments we are able to determine the unknown refractive indices n_1 and n_3 as well as the inclination angle θ of the indicatrix in the ac-plane with respect to the normal of the ab-plane (see Fig. 3.1). Using the standard equation for the description of the refractive indices experienced by a light wave with arbitrary propagation direction in a biaxial crystal and the determined value of n_{13}, n_{TM} and ϕ' at a wavelength of 1.064 µm we are able to determine the orientation and the shape of the indicatrix. The main refractive indices and the inclination angle are given by $n_1 = 2.5 \pm 0.2$, $n_2 = 1.60 \pm 0.02$, $n_3 = 1.60 \pm 0.03$ and $\theta = 39 \pm 3°$ at a wavelength of 1.064 µm. The relatively large error in n_1 compared to n_2 or n_3 is due to an unfavorable inclination angle of the indicatrix for the evaluation of the data. However, in the analysis of the electro-optic measurements we mainly make use of n_{TM}, which was determined with a considerably better relative accuracy.

The results show that DAT2 is highly anisotropic with a birefringence of $\Delta n = n_1 - n_2 = n_1 - n_3 = 0.9 \pm 0.2$ at $\lambda = 1.064$ µm both in the ac crystallographic plane and in the plane perpendicular to it containing x_1. This is more than 40% larger than in DAST, for which $\Delta n \simeq 0.62$ at $\lambda = 1.064$ µm [67]. The large birefringence Δn can be well understood by the highly aligned chromophores combined with the large electron polarizability along the long axis of the chromophores compared to the one in the plane perpendicular to x_1 (see Fig. 3.1). The very similar refractive index values of x_2 and x_3 can be explained by an angle of roughly 90° enclosed by the molecular planes of the two chromophores in the unit cell and the crystallographic b-axis as bisecting line of this angle.

APPENDIX C

Appendix to chapter 4

C.1 DAT2 Mach-Zehnder modulators

The fabrication technique we developed, in which the melt of the organic material flows into predefined channels by capillary force and crystallizes there upon cooling (see chapter 3), allowed for the fabrication of high-quality single-crystalline phase modulators of DAT2 as well as of electro-optic microring resonators of COANP. Here we illustrate the versatility of this fabrication technique, by the production of electro-optic crystalline Mach-Zehnder modulators based on DAT2.

Material and crystal structure

The monoclinic DAT2 crystals, with point group symmetry 2, have the twofold symmetry axis along the polar b axis. The main charge transfer axis of the chromophore is tilted by an angle of about 56° with respect to the largest thin film crystal face, which is parallel to the ab-crystallographic plane. The details of the crystal orientation and the linear optical properties have been reported in chapter 3.

Sample production

In this section the technique of fabricating electro-optic crystalline Mach-Zehnder modulators of DAT2 is described. The basic concept is the same as for the microring resonators introduced in chapter 4. However, the production of Mach-Zehnder modulator samples is somewhat more complicated compared to microring resonator structures, since both the substrate as well as the cover wafer had to be photolithographically processed, and finally precise alignment of the cover to the substrate wafer was required.

We used standard borosilicate glass wafers for the substrate and cover material. On the substrate wafer 70 nm chromium and 50 nm amorphous silicon were deposited by electron-gun evaporation and patterned by conventional photolithographic processing and a standard lift-off technique (see Fig. C1a). The chromium was used as electrical contacts to the final Mach-Zehnder modulator and silicon was required for anodic bonding. Spinning of a roughly 180 nm thick layer of a spin-on glass (SOG) solution (Filmtronics 700A) onto the electroded substrate was of avail to reduce reflections and guiding losses in the regions where the electrodes cross the waveguides, which was required in order to contact the inside of the Mach-Zehnder structures. In addition SOG is regularly used for adhesive wafer bonding techniques [134] and also in our configuration it acted as an adhesive for the alignment and bonding process described below.

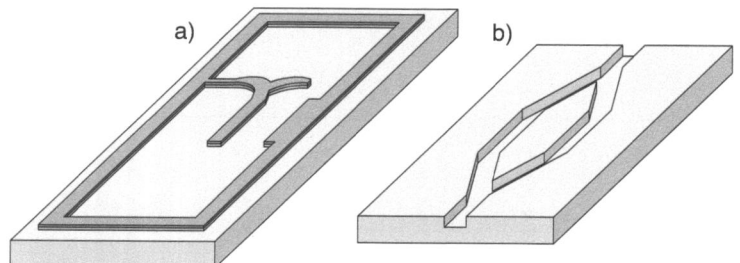

Figure C1: Processing steps for the fabrication of Mach-Zehnder modulators. a) 70 nm chromium and 50 nm amorphous silicon were deposited and structured on a borosilicate substrate wafer. The chromium was used to apply an electric field over one arm of the Mach-Zehnder modulator and the silicon was required for anodic bonding. b) The Mach-Zehnder waveguide structure was patterned by RIE into a cover borosilicate glass. After spinning an SOG solution onto the substrate, the cover was aligned and bonded to the substrate. A microscope image of the final device is shown in Fig. C2.

The Mach-Zehnder waveguide configuration was structured into the cover wafer analogously as in the case of microring resonators. By fabricating the waveguide pattern in the glass, the accessibility of the waveguide pattern for the melt of the organic material from the edges of the cover glass had to be ensured as illustrated in Fig. C1b. Due to the commonly applied photoresist edge bead removal and long enough straight waveguide sections, the Mach-Zehnder configuration was consequently accessible for the melt. The structured cover borosilicate was then aligned to the electroded substrate wafer in a home-built alignment setup and held in place by the adhesion due to the SOG on the substrate. Finally an anodic bonding step was applied to bond the two wafers permanently together (the bonding parameters have been reported in chapter 3).

Crystal growth

The DAT2 crystals were grown by a variation of the Bridgman method reported in chapter 3. A fabricated Mach-Zehnder modulator with the organic crystalline DAT2 material grown in the waveguide channels is shown in Fig. C2. As for the phase modulators reported in chapter 3 the DAT2 material in the electro-optically tuned Mach-Zehnder arm was found to crystallize with the polar *b* axis perpendicular to the waveguide channel, which is highly favorable for applying the electro-optic effect.

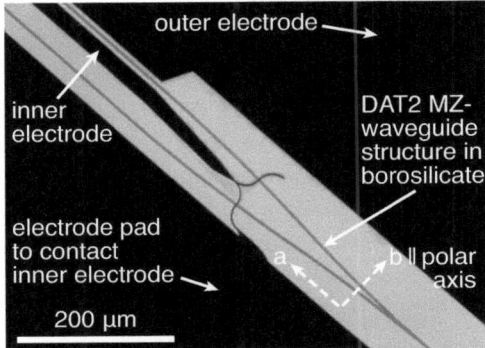

Figure C2: Top view of a fabricated single-crystalline DAT2 Mach-Zehnder device. The organic crystals of DAT2 were grown from the melt in the waveguide channels after aligning and anodically bonding the borosilicate cover wafer (where the waveguide structure was patterned) to the substrate wafer (where the metal electrodes were deposited). The bright areas adjacent to the waveguides present an approximately 120 nm thin DAT2 crystalline layer.

Characterization of electro-optic Mach-Zehnder modulators

The integrated single-crystalline DAT2 Mach-Zehnder interferometers were operated by launching either transverse magnetic (TM) or transverse electric (TE) modes at the telecommunication wavelength 1.55 μm in the waveguide. Modulation experiments were performed by applying a sinusoidal electric signal to the electrodes and measuring the resulting light intensity modulation. An alternating current transformer was used to increase the voltage produced by a conventional 10 V signal generator. An oscilloscope was used to record the applied voltage $V(t)$ and the output light intensity. The behavior of the signal intensity at the output of the Mach-Zehnder modulator is described by its switching curve, which can be written as a function of the applied voltage $V(t)$ as

$$I(t) = \frac{1}{2}\left(I_1 + I_2 + 2 \cdot \sqrt{I_1 I_2} \cdot \cos\left(\phi_0 + \frac{\pi V(t)}{V_\pi}\right)\right) \quad \text{(C1)}$$

with the electro-optic phase shift $\Delta\phi(t) = \pi V(t)/V_\pi$, where V_π is the half-wave voltage, defined as the voltage required to induce a phase shift of π between the two arms of the interferometer. ϕ_0 is the initial phase shift resulting from imperfect symmetry of the two arms and I_1 and I_2 are the intensities in the two arms. By substituting $V(t)$ in Eq. (C1) with a sinusoidal voltage $V(t) = V_m \sin(\omega_m t)$ as used in the experiment, we obtain for the peak-to-peak value of the modulated signal intensity

$$\Delta I = 4 \cdot \sqrt{I_1 I_2} \cdot \sin(\phi_0) \sin\left(\frac{\pi V_m}{V_\pi}\right) = \Delta I_0 \sin\left(\frac{\pi V_m}{V_\pi}\right), \quad \text{(C2)}$$

C. Appendix to chapter 4

which is valid as long as the modulation voltage V_m is small enough not to generate more than two extremal points (one minimum and maximum) in the modulated signal per oscillation period of the applied sinusoidal voltage; i.e. for $V_m/V_\pi < |\phi_0 \bmod \pi|/\pi$ and $V_m/V_\pi < 1-|\phi_0 \bmod \pi|/\pi$ (for $\phi_0 = \pi/2$ this simplifies to $V_m < V_\pi/2$). Several peak-to-peak signal intensities corresponding to different amplitudes of the modulation voltage were recorded with the help of an oscilloscope and are shown in Fig. C3 for TE- and TM-modes. The solid curves in Fig. C3 were obtained from a least square theoretical analysis of the model in Eq. (C2) to the data $\Delta I(V_m)$ from the varying modulation amplitude voltage measurement. Like this, we could determine the half-wave voltage V_π, while all the unknown parameters (I_1, I_2 and ϕ_0) only appear in the normalization constant ΔI_0. The obtained half-wave voltage of the single-arm phase-retardation Mach-Zehnder interferometer with an interaction length of 2 mm results in a half-wave voltage × length product of $V_\pi L = 78 \pm 2$ Vcm for TE-modes and $V_\pi L = 60 \pm 1$ Vcm for TM-modes at 1.55 µm.

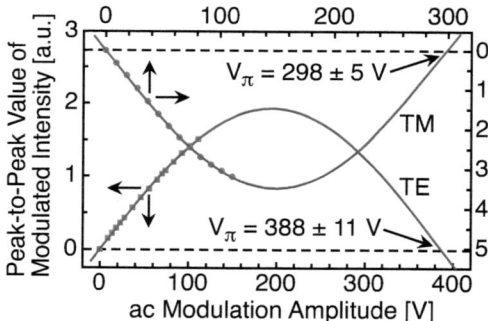

Figure C3: Peak-to-peak modulated light intensity ΔI detected by a photodiode as a function of the amplitude of the sinusoidal modulation voltage V_m applied to the side electrodes. The crystalline DAT2 Mach-Zehnder interferometers with an electric field applied over only one 2 mm long arm featured a half-wave voltage of $V_\pi = 388 \pm 11$ V for TE-modes and $V_\pi = 298 \pm 5$ V for TM-modes.

The experimentally determined half-wave voltages allow to calculate the linear electro-optic tensor elements r_{ijk} relevant for the field-induced refractive index change. Due to the crystal orientation the electro-optic modulation measurements were performed with mode propagation along the crystallographic a-axis. Because of an inclined indicatrix in the ac-plane several tensor elements r_{ijk} contribute to the phase shift for TM modes in the Mach-Zehnder waveguide. However, we have previously argued [121], why the phase shift for TM modes in the waveguide is mainly caused by r_{112} and why changes induced by other tensor elements may be neglected. Also due to the inclined indicatrix we have to consider the projection factors of the dielectric system in which r_{ijk} is given onto the reference frame of a TM mode. Taking this into account the electro-optic coefficient r_{112} is given by

$$r_{112} = \frac{1}{V_\pi} \cdot \frac{1}{(\cos\theta)^2} \cdot \frac{\lambda d}{l n_{\text{eff}} n^2}, \tag{C3}$$

where V_π is the measured half-wave voltage, θ is the angle between the dielectric axis x_1 and the normal to the crystallographic ab-plane, l is the interaction length of applied electric field and the organic crystal waveguide, n_{eff} is the effective refractive index experienced by the waveguide mode and n is the bulk refractive index of the chosen polarization. In Eq. (C3) both n and n_{eff} appear since in a first order approximation the same proportionality constant was used to relate n_{eff} and n as to relate the change of the corresponding refractive indices due to the electro-optic effect, i.e. Δn_{eff} and Δn. The bulk refractive index value n as well as the inclination angle θ for DAT2 crystals have been reported in literature [121], i.e. $n^{\text{TM}} = 1.995$, $n^{\text{TE}} = n_2 = 1.594$ and $\theta = 39°$ at 1.55 µm. The effective refractive index for TM modes in the waveguide with cross-sectional dimensions of $w \times h = 2.66 \times 1.3$ µm^2 was calculated with the full vectorial complex FD Generic mode solver of the commercially available software OlympIOs [110] ($n_{\text{eff}}^{\text{TM}} = 1.91$, for simplicity considering the isotropic refractive index n^{TM} of the waveguide). Using these values and Eq. (C3) the electro-optic coefficient $r_{112} = 7.4 \pm 0.4$ pm/V was obtained at a wavelength of 1.55 µm and this value is within experimental error in good agreement with the one previously measured in DAT2 phase modulators [121].

With exactly the same approach also the value of r_{222} was determined, which causes the phase shift for TE modes. However, since the dielectric axis x_2 is perpendicular to the waveguide channel the projection angle θ in Eq. (C3) is zero for TE modes and with $n_{\text{eff}}^{\text{TE}} = 1.53$ at 1.55 µm we obtain $r_{222} = 6.7 \pm 0.4$ pm/V at 1.55 µm.

Conclusion

Due to the versatility of the fabrication method introduced in chapter 3 we were also able to fabricate electro-optic Mach-Zehnder modulators based on crystalline DAT2 material. DAT2 was chosen since it crystallizes with the polar b axis perpendicular to the waveguide channels. The resultant device with a 2 mm long interferometer arm was measured with 1.55 µm incident light. Electro-optic modulation has been demonstrated and the half-wave voltage × length product could be estimated to be $V_\pi L = 78 \pm 2$ Vcm for TE-modes and $V_\pi L = 60 \pm 1$ Vcm for TM-modes, where the latter is in good agreement with earlier measurements in phase modulators.

C. Appendix to chapter 4

Bibliography

[1] A. G. Bell, "Selenium and the photophone," Nature **22**, 500–503 (1880).

[2] TOP500 Supercomputer Sites, "TOP500 list - November 2008," (2008). "Available at http://www.top500.org".

[3] Intel News Release, "Intel accelerates high performance computing clusters," (2007). "Available at http://www.intel.com/pressroom/archive/releases/20070627corp.htm".

[4] Barcelona Supercomputing Center, "MareNostrum," (2009). "Available at http://www.bsc.org.es".

[5] M. Paniccia, L. Liao, A. Liu, H. Rong, and S. M. Koehl, "Silicon-integrated optics," in "Optical Interconnects," , vol. 119, L. Pavesi and G. Guillot, eds. (Springer-Verlag, Berlin, 2006), chap. 12.

[6] S. Libertino and A. Sciuto, "Electro-optical modulators in silicon," in "Optical Interconnects," , vol. 119, L. Pavesi and G. Guillot, eds. (Springer-Verlag, Berlin, 2006), chap. 4.

[7] B. Jalali, "Can silicon change photonics?" phys. stat. sol. (a) **205**, 213–224 (2008).

[8] A. Karlsson, R. Schatz, and G. Bjork, "On the modulation bandwidth of semiconductor microcavity lasers," IEEE Photon. Technol. Lett. **6**, 1312–1314 (1994).

[9] J. Liu, M. Beals, A. Pomerene, S. Bernardis, R. Sun, J. Cheng, L. C. Kimerling, and J. Michel, "Waveguide-integrated, ultralow-energy GeSi electro-absorption modulators," Nat. Photon. **2**, 433–437 (2008).

[10] Covega Corporation, "40 Gb/s lithium niobate modulators," (2008). "Available at http://www.covega.com/".

[11] Avanex Corporation, "Lithium niobate for 40G modulation," (2008). "Available at http://www.avanex.com/".

BIBLIOGRAPHY

[12] Q. Xu, B. Schmidt, S. Pradhan, and M. Lipson, "Micrometre-scale silicon electro-optic modulator," Nature **435**, 325–327 (2005).

[13] Q. Xu, S. Manipatruni, B. Schmidt, J. Shakya, and M. Lipson, "12.5 Gbit/s carrier-injection-based silicon micro-ring silicon modulators," Opt. Express **15**, 430–436 (2007).

[14] L. Liao, A. Liu, D. Rubin, J. Basak, Y. Chetrit, H. Nguyen, R. Cohen, N. Izhaky, and M. Paniccia, "40 Gbit/s silicon optical modulator for high-speed applications," Electron. Lett. **43**, 1196–1197 (2007).

[15] D. Chen, H. R. Fetterman, A. Chen, W. H. Steier, L. R. Dalton, W. Wang, and Y. Shi, "Demonstration of 110 GHz electro-optic polymer modulators," Appl. Phys. Lett. **70**, 3335–3337 (1997).

[16] Y. Shi, C. Zhang, H. Zhang, J. H. Bechtel, L. R. Dalton, B. H. Robinson, and W. H. Steier, "Low (sub-1-volt) halfwave voltage polymeric electro-optic modulators achieved by controlling chromophore shape," Science **288**, 119–122 (2000).

[17] B. Bortnik, Y.-C. Hung, H. Tazawa, B.-J. Seo, J. Luo, A. K.-Y. Jen, W. H. Steier, and H. R. Fetterman, "Electrooptic polymer ring resonator modulation up to 165 GHz," IEEE J. Sel. Top. Quantum Electron. **13**, 104–110 (2007).

[18] Y. Enami, D. Mathine, C. T. DeRose, R. A. Norwood, J. Luo, A. K.-Y. Jen, and N. Peyghambarian, "Hybrid cross-linkable polymer/sol-gel waveguide modulators with 0.65 V half wave voltage at 1550 nm," Appl. Phys. Lett. **91**, 093505 (2007).

[19] M. Lee, H. E. Katz, C. Erben, D. M. Gill, P. Gopalan, J. D. Heber, and D. J. McGee, "Broadband modulation of light by using an electro-optic polymer," Science **298**, 1401–1403 (2002).

[20] B. M. A. Rahman, S. Haxha, V. Haxha, and K. T. V. Grattan, "Design optimization of high-speed optical modulators," Proc. SPIE **6389**, 63890X (2006).

[21] Y. Enami, C. T. Derose, D. Mathine, C. Loychik, C. Greenlee, R. A. Norwood, T. D. Kim, J. Luo, Y. Tian, A. K.-Y. Jen, and N. Peyghambarian, "Hybrid polymer/sol-gel waveguide modulators with exceptionally large electro-optic coefficients," Nat. Photon. **1**, 180–185 (2007).

[22] V. R. Almeida, Q. Xu, C. A. Barrios, and M. Lipson, "Guiding and confining light in void nanostructure," Opt. Lett. **29**, 1209–1211 (2004).

[23] J.-M. Brosi, C. Koos, L. C. Andreani, M. Waldow, J. Leuthold, and W. Freude, "High-speed low-voltage electro-optic modulator with a polymer-infiltrated silicon photonic crystal waveguide," Opt. Express **16**, 4177–4191 (2008).

[24] M. Hochberg, T. Baehr-Jones, G. Wang, J. Huang, P. Sullivan, L. Dalton, and A. Scherer, "Towards a millivolt optical modulator with nano-slot waveguides," Opt. Express **15**, 8401–8410 (2007).

[25] T. Baehr-Jones, B. Penkov, J. Huang, P. Sullivan, J. Davies, J. Takayesu, J. Luo, T.-D. Kim, L. Dalton, A. Jen, M. Hochberg, and A. Scherer, "Nonlinear polymer-clad silicon slot waveguide modulator with a half wave voltage of 0.25 V," Appl. Phys. Lett. **92**, 163303 (2008).

[26] B. E. A. Saleh and M. C. Teich, *Fundamentals of Photonics* (John Wiley and Sons, Inc., New York, 1991).

[27] P. N. Butcher and D. Cotter, *The Elements of Nonlinear Optics* (Cambridge University Press, Cambridge, 1990).

[28] C. Bosshard, K. Sutter, P. Prêtre, J. Hulliger, M. Flörsheimer, P. Kaatz, and P. Günter, *Organic Nonlinear Optical Materials* (Gordon and Breach, Basel, 1995).

[29] F. Heismann, S. K. Korotky, and J. J. Veselka, "Lithium niobate integrated optics: Selected contemporary devices and system applications," in "Optical Fiber Telecommunications IIIB," , vol. IIIB, I. P. Kaminow and T. L. Koch, eds. (Academic Press, San Diego, 1997), chap. 9.

[30] S.-J. Chang, C.-L. Tsai, Y.-B. Lin, J.-F. Liu, and W.-S. Wang, "Improved electrooptic modulator with ridge structure in x-cut $LiNbO_3$," J. Lightwave Technol. **17**, 843–847 (1999).

[31] F. Michelotti, A. Driessen, and M. Bertolotti, *Microresonators as Building Blocks for VLSI Photonics* (AIP Conference Proceedings, Melville, New York, 2004).

[32] J. Heebner, R. Grover, and T. A. Ibrahim, *Optical Microresonators* (Springer-Verlag, London, 2008).

[33] J. H. Wülbern, A. Petrov, and M. Eich, "Electro-optical modulator in a polymerinfiltrated silicon slotted photonic crystal waveguide heterostructure resonator," Opt. Express **17**, 304–313 (2009).

[34] A. Guarino, G. Poberaj, D. Rezzonico, R. Degl'Innocenti, and P. Günter, "Electro-optically tunable microring resonators in lithium niobate," Nat. Photon. **1**, 407–410 (2007).

[35] P. Rabiei, W. H. Steier, C. Zhang, and L. R. Dalton, "Polymer micro-ring filters and modulators," J. Lightwave Technol. **20**, 1968–1975 (2002).

[36] D. Rezzonico, M. Jazbinšek, A. Guarino, O.-P. Kwon, and P. Günter, "Electro-optic Charon polymeric microring modulators," Opt. Express **16**, 613–627 (2008).

BIBLIOGRAPHY

[37] H. Sun, A. Chen, B. C. Olbricht, J. A. Davies, P. A. Sullivan, Y. Liao, and L. R. Dalton, "Direct electron beam writing of electro-optic polymer microring resonators," Opt. Express **16**, 6592–6599 (2008).

[38] Q. Xu, B. Schmidt, J. Shakya, and M. Lipson, "Cascaded silicon micro-ring modulators for WDM optical interconnection," Opt. Express **14**, 9431–9435 (2006).

[39] I.-L. Gheorma and R. M. Osgood, "Fundamental limitations of optical resonator based high-speed EO modulators," IEEE Photon. Tech. Lett. **14**, 795–797 (2002).

[40] A. Guarino, "Electro-optic microring resonators in inorganic crystals for photonic applications," Ph.D. thesis, ETH Zürich (2007).

[41] P. J. Winzer and R.-J. Essiambre, "Advanced modulation formats for high-capacity optical transport networks," J. Lightwave Technol. **24**, 4711–4728 (2006).

[42] T. Tokle, M. Serbay, J. B. Jensen, W. Rosenkranz, and P. Jeppesen, "Advanced modulation formats for transmission systems," OFC Conference 2008 **OMI1** (2008).

[43] M. Seimetz, "Optical fiber transmission systems with high-order phase and quadrature amplitude modulation," Ph.D. thesis, Techn. Univ. Berlin (2008).

[44] X. Zhou, J. Yu, M.-F. Huang, Y. Shao, T. Wang, P. Magill, M. Cvijetic, L. Nelson, M. Birk, G. Zhang, S. Y. Ten, H. B. Matthew, and S. K. Mishra, "32 Tb/s (320x114 Gb/s) PDM-RZ-8QAM transmission over 580 km of SMF-28 ultra-low-loss fiber," OFC Conference 2009 **PDPB4** (2009).

[45] H. Masuda, E. Yamazaki, A. Sano, T. Yoshimatsu, T. Kobayashi, E. Yoshida, Y. Miyamoto, S. Matsuoka, Y. Takatori, M. Mizoguchi, K. Okada, K. Hagimoto, T. Yamada, and S. Kamei, "13.5-Tb/s (135 x 111-Gb/s/ch) no-guard-interval coherent OFDM transmission over 6,248 km using SNR maximized second-order DRA in the extended L-band," OFC Conference 2009 **PDPB5** (2009).

[46] H. Takahashi, A. A. Amin, S. L. Jansen, I. Morita, and H. Tanaka, "DWDM transmission with 7.0-bit/s/Hz spectral efficiency using 8x65.1-Gbit/s coherent PDM-OFDM signals," OFC Conference 2009 **PDPB7** (2009).

[47] H. G. Weber, S. Ferber, M. Kroh, C. Schmidt-Langhorst, R. Ludwig, V. Marembert, C. Boerner, F. Futami, S. Watanabe, and C. Schubert, "Single channel 1.28 Tbit/s and 2.56 Tbit/s DQPSK transmission," Electron. Lett. **42**, 178–179 (2006).

[48] C. Schmidt-Langhorst, R. Ludwig, D.-D. Groß, L. Molle, M. Seimetz, R. Freund, and C. Schubert, "Generation and coherent time-division demultiplexing of up to 5.1 Tb/s single-channel 8-PSK and 16-QAM signals," OFC Conference 2009 **PDPC6** (2009).

[49] P. J. Winzer, C. Dorrer, R.-J. Essiambre, and I. Kang, "Chirped return-to-zero modulation by imbalanced pulse carver driving signals," IEEE Photon. Technol. Lett. **16**, 1379–1381 (2004).

[50] A. H. Gnauck, G. Charlet, P. Tran, P. J. Winzer, C. R. Doerr, J. C. Centanni, E. C. Burrows, T. Kawanishi, T. Sakamoto, and K. Higuma, "25.6-Tb/s WDM transmission of polarization-multiplexed RZ-DQPSK signals," J. Lightwave Technol. **26**, 79–84 (2008).

[51] T. Kawanishi, T. Sakamoto, and M. Izutsu, "High-speed control of lightwave amplitude, phase, and frequency by use of electrooptic effect," IEEE J. Sel. Top. Quantum Electron. **13**, 79–91 (2007).

[52] G. Charlet, "Coherent detection associated with digital signal processing for fiber optics communication," C. R. Phys. **9**, 1012 – 1030 (2008).

[53] ITF Laboratories Inc., "Phase demodulators," (2009). "Available at http://www.itflabs.com".

[54] Optoplex Corporation, "Optical demodulators," (2009). "Available at http://www.optoplex.com".

[55] M. Seimetz and C.-M. Weinert, "Options, feasibility, and availability of 2 × 4 90° hybrids for coherent optical systems," J. Lightwave Technol. **24**, 1317–1322 (2006).

[56] L. Zhang, J.-Y. Yang, M. Song, Y. Li, B. Zhang, R. G. Beausoleil, and A. E. Willner, "Microring-based modulation and demodulation of DPSK signal," Opt. Express **15**, 11564–11569 (2007).

[57] L. Zhang, J.-Y. Yang, Y. Li, M. Song, R. G. Beausoleil, and A. E. Willner, "Monolithic modulator and demodulator of differential quadrature phase-shift keying signals based on silicon microrings," Opt. Lett. **33**, 1428–1430 (2008).

[58] Y. Chen and S. Blair, "Nonlinear phase shift of cascaded microring resonators," J. Opt. Soc. Am. B **20**, 2125–2132 (2003).

[59] H. Ma, A. K.-Y. Jen, and L. R. Dalton, "Polymer-based optical waveguides: Materials, processing, and devices," Adv. Mater. **14**, 1339–1365 (2002).

[60] O.-P. Kwon, B. Ruiz, A. Choubey, L. Mutter, A. Schneider, M. Jazbinšek, V. Gramlich, and P. Günter, "Organic nonlinear optical crystals based on configurationally locked polyene for melt growth," Chem. Mater. **18**, 4049–4054 (2006).

[61] H. S. Nalwa, T. Watanabe, and S. Miyata, "Organic materials for second-order nonlinear optics," in "Nonlinear Optics of Organic Molecules and Polymers," , H. S. Nalwa and S. Miyata, eds. (CRC Press, Boca Raton, Florida, 1997), chap. 4.

BIBLIOGRAPHY

[62] C. Bosshard, M. Bösch, I. Liakatas, M. Jäger, and P. Günter, "Second-order nonlinear optical organic materials: Recent developments," in "Nonlinear Optical Effects and Materials,", vol. 72, P. Günter, ed. (Springer-Verlag, Berlin, 2000), chap. 3.

[63] M. Jazbinsek, L. Mutter, and P. Günter, "Photonic applications with the organic nonlinear optical crystal DAST," IEEE J. Sel. Top. Quantum Electron. **14**, 1298–1311 (2008).

[64] S. R. Marder, J. W. Perry, and W. P. Schaefer, "Synthesis of organic salts with large second-order optical nonlinearities," Science **245**, 626–628 (1989).

[65] S. R. Marder, J. W. Perry, and C. P. Yakymyshyn, "Organic salts with large second-order optical nonlinearities," Chem. Mater. **6**, 1137–1147 (1994).

[66] U. Meier, M. Bösch, C. Bosshard, F. Pan, and P. Günter, "Parametric interactions in the organic salt 4-N, N-dimethylamino-4'-N'-methyl-stilbazolium tosylate at telecommunication wavelengths," J. Appl. Phys. **83**, 3486–3489 (1998).

[67] F. Pan, G. Knöpfle, C. Bosshard, S. Follonier, R. Spreiter, M. S. Wong, and P. Günter, "Electro-optic properties of the organic salt 4-N, N-dimethylamino-4'-N'-methyl-stilbazolium tosylate," Appl. Phys. Lett. **69**, 13–15 (1996).

[68] A. Schneider, M. Neis, M. Stillhart, B. Ruiz, R. U. A. Khan, and P. Günter, "Generation of terahertz pulses through optical rectification in organic DAST crystals: Theory and experiment," J. Opt. Soc. Am. B **23**, 1822–1835 (2006).

[69] T. Taniuchi, S. Ikeda, S. Okada, and H. Nakanishi, "Tunable sub-terahertz wave generation from an organic DAST crystal," Jap. J. Appl. Phys. **44**, L652–L654 (2005).

[70] T. Kaino, B. Cai, and K. Takayama, "Fabrication of DAST channel optical waveguides," Adv. Funct. Mater. **12**, 599–603 (2002).

[71] W. Geis, R. Sinta, W. Mowers, S. J. Deneault, M. F. Marchant, K. E. Krohn, S. J. Spector, D. R. Calawa, and T. M. Lyszczarz, "Fabrication of crystalline organic waveguides with an exceptionally large electro-optic coefficient," Appl. Phys. Lett. **84**, 3729–3731 (2004).

[72] L. Mutter, A. Guarino, M. Jazbinsek, M. Zgonik, P. Günter, and M. Döbeli, "Ion implanted optical waveguides in nonlinear optical organic crystal," Opt. Express **15**, 629–638 (2007).

[73] L. Mutter, M. Koechlin, M. Jazbinsek, and P. Günter, "Direct electron beam writing of channel waveguides in nonlinear optical organic crystals," Opt. Express **15**, 16828–16838 (2007).

[74] Z. Yang, S. Aravazhi, P. Seiler, M. Jazbinsek, and P. Günter, "Synthesis and crystal growth of stilbazolium derivatives for second-order nonlinear optics," Adv. Funct. Mater. **15**, 1072–1076 (2005).

[75] Z. Glavcheva, H. Umezawa, Y. Mineno, T. Odani, S. Okada, S. Ikeda, T. Taniuchi, and H. Nakanishi, "Synthesis and properties of 1-methyl-4-{2-[4-(dimethylamino)phenyl]ethenyl}pyridinium p-toluenesulfonate derivatives with isomorphous crystal structure," Jpn. J. Appl. Phys., Part 1 **44**, 5231–5235 (2005).

[76] B. Ruiz, Z. Yang, V. Gramlich, M. Jazbinšek, and P. Günter, "Synthesis and crystal structure of a new stilbazolium salt with large second-order optical nonlinearity," J. Mater. Chem. **16**, 2839 (2006).

[77] Z. Yang, M. Jazbinšek, B. Ruiz, S. Aravazhi, V. Gramlich, and P. Günter, "Molecular engineering of stilbazolium derivatives for second-order nonlinear optics," Chem. Mater. **19**, 3512–3518 (2007).

[78] Z. Yang, L. Mutter, M. Stillhart, B. Ruiz, S. Aravazhi, M. Jazbinšek, A. Schneider, V. Gramlich, and P. Günter, "Large-size bulk and thin-film stilbazolium-salt single crystals for nonlinear optics and THz generation," Adv. Funct. Mater. **17**, 2018–2023 (2007).

[79] L. Mutter, F. D. Brunner, Z. Yang, M. Jazbinšek, and P. Günter, "Linear and nonlinear optical properties of the organic crystal DSTMS," J. Opt. Soc. Am. B **24**, 2556–2561 (2007).

[80] B. J. Coe, J. P. Essex-Lopresti, J. A. Harris, S. Houbrechts, and A. Persoons, "Ruthenium(II) ammine centres as efficient electron donor groups for quadratic non-linear optics," Chem. Commun. pp. 1645–1646 (1997).

[81] B. J. Coe, J. A. Harris, L. J. Harrington, J. C. Jeffery, L. H. Rees, S. Houbrechts, and A. Persoons, "Enhancement of molecular quadratic hyperpolarizabilities in Ruthenium(II) 4,4'-bipyridinium complexes by N-phenylation," Inorg. Chem. **37**, 3391–3399 (1998).

[82] B. J. Coe, J. A. Harris, I. Asselberghs, K. Clays, G. Olbrechts, A. Persoons, J. T. Hupp, R. C. Johnson, S. J. Coles, M. B. Hursthouse, and K. Nakatani, "Quadratic nonlinear optical properties of N-aryl stilbazolium dyes," Adv. Funct. Mater. **12**, 110–116 (2002).

[83] B. J. Coe, J. A. Harris, I. Asselberghs, K. Wostyn, K. Clays, A. Persoons, B. S. Brunschwig, S. J. Coles, T. Gelbrich, M. E. Light, M. B. Hursthouse, and K. Nakatani, "Quadratic optical nonlinearities of N-methyl and N-aryl pyridinium salts," Adv. Funct. Mater. **13**, 347–357 (2003).

[84] B. J. Coe, D. Beljonne, H. Vogel, J. Garin, and J. Orduna, "Theoretical analyses of the effects on the linear and quadratic nonlinear optical properties of N-arylation of pyridinium groups in stilbazolium dyes," J. Phys. Chem. A **109**, 10052–10057 (2005).

BIBLIOGRAPHY

[85] B. Ruiz, B. J. Coe, R. Gianotti, V. Gramlich, M. Jazbinšek, and P. Günter, "Polymorphism, crystal growth and characterization of an organic nonlinear optical material: DAPSH," CrystEngComm **9**, 772–776 (2007).

[86] M. S. Shumate, "Interferometric measurements of large indices of refraction," Appl. Opt. **5**, 327–331 (1966).

[87] J. G. Bergman and G. R. Crane, "Structural aspects of nonlinear optics: Optical properties of KIO_2F_2 and its related iodates," J. Chem. Phys. **60**, 2470–2474 (1974).

[88] J. Zyss and J. L. Oudar, "Relations between microscopic and macroscopic lowest-order optical nonlinearities of molecular crystals with one- or two-dimensional units," Phys. Rev. A **26**, 2028–2048 (1982).

[89] J. Jerphagnon and S. K. Kurtz, "Maker fringes: A detailed comparison of theory and experiment for isotropic and uniaxial crystals," J. Appl. Phys. **41**, 1667–1681 (1970).

[90] R. C. Miller, "Optical second harmonic generation in piezoelectric crystals," Appl. Phys. Lett. **5**, 17–19 (1964).

[91] I. Liakatas, C. Cai, M. Bosch, M. Jager, C. Bosshard, P. Günter, C. Zhang, and L. R. Dalton, "Importance of intermolecular interactions in the nonlinear optical properties of poled polymers," Appl. Phys. Lett. **76**, 1368–1370 (2000).

[92] M. Schmidt, M. Eich, U. Huebner, and R. Boucher, "Electro-optically tunable photonic crystals," Appl. Phys. Lett. **87**, 121110 (2005).

[93] L. R. Dalton, P. A. Sullivan, D. H. Bale, and B. C. Olbricht, "Theory-inspired nano-engineering of photonic and electronic materials: Noncentrosymmetric charge-transfer electro-optic materials," Solid-State Electron. **51**, 1263–1277 (2007).

[94] D. Rezzonico, S.-J. Kwon, H. Figi, O.-P. Kwon, M. Jazbinšek, and P. Günter, "Photochemical stability of nonlinear optical chromophores in polymeric and crystalline materials," J. Chem. Phys. **128**, 124713 (2008).

[95] S. Manetta, M. Ehrensperger, C. Bosshard, and P. Günter, "Organic thin film crystal growth for nonlinear optics: Present methods and exploratory developments," Comptes Rendus Physique **3**, 449–462 (2002).

[96] M. Thakur, J. Titus, and A. Mishra, "Single-crystal thin films of organic molecular salt may lead to a new generation of electro-optic devices," Opt. Eng. **42**, 456–458 (2003).

[97] L. Mutter, M. Jazbinšek, M. Zgonik, U. Meier, C. Bosshard, and P. Günter, "Photobleaching and optical properties of organic crystal 4-N, N-dimethylamino-4'-N'-methyl-stilbazolium tosylate," J. Appl. Phys. **94**, 1356–1361 (2003).

BIBLIOGRAPHY

[98] P. Dittrich, R. Bartlome, G. Montemezzani, and P. Günter, "Femtosecond laser ablation of DAST," Appl. Surface Science **220**, 88–95 (2003).

[99] L. Mutter, M. Jazbinšek, C. Herzog, and P. Günter, "Electro-optic and nonlinear optical properties of ion implanted waveguides in organic crystals," Opt. Express **16**, 731–739 (2008).

[100] S. Gauvin and J. Zyss, "Growth of organic crystalline thin films, their optical characterization and application to non-linear optics," J. Cryst. Growth **166**, 507–527 (1996).

[101] A. Leyderman, Y. Cui, and B. G. Penn, "Electro-optical effects in thin single-crystalline organic films grown from the melt," J. Phys. D: Appl. Phys. **31**, 2711–2717 (1998).

[102] A. Choubey, O.-P. Kwon, M. Jazbinšek, and P. Günter, "High-quality organic single crystalline thin films for nonlinear optical applications by vapor growth," Cryst. Growth Des. **7**, 402–405 (2007).

[103] O.-P. Kwon, S.-J. Kwon, H. Figi, M. Jazbinšek, and P. Günter, "Organic electro-optic single-crystalline thin films grown directly on modified amorphous substrates," Adv. Mater. **20**, 543–545 (2008).

[104] M. A. Schmidt, "Wafer-to-wafer bonding for microstructure formation," Proc. IEEE **86**, 1575–1585 (1998).

[105] Q.-Y. Tong and U. Gösele, *Semiconductor Wafer Bonding: Science and Technology* (John Wiley & Sons, New York, 1999).

[106] P. Lindner, V. Dragoi, S. Farrens, T. Glinsner, and P. Hangweier, "Advanced techniques for 3D devices in wafer-bonding processes," Solide State Technol. **47**, 55–58 (2004).

[107] A. Berthold, L. Nicola, P. M. Sarro, and M. J. Vellekoop, "Glass-to-glass anodic bonding with standard IC technology thin films as intermediate layers," Sens. Actuators, A **82**, 224–228 (2000).

[108] P. V. Vidakovic, M. Coquillay, and F. Salin, "N-(4-nitrophenyl)-N-methylamino-acetonitrile: A new organic material for efficient second-harmonic generation in bulk and waveguide configurations. I. Growth, crystal structure, and characterization of organic crystal-cored fibers," J. Opt. Soc. Am. B **4**, 998–1012 (1987).

[109] S.-J. Kwon, O.-P. Kwon, J.-I. Seo, M. Jazbinšek, L. Mutter, V. Gramlich, Y.-S. Lee, H. Yun, and P. Günter, "Highly nonlinear optical configurationally locked triene crystals based on 3,5-dimethyl-2-cyclohexen-1-one," J. Phys. Chem. C **112**, 7846–7852 (2008).

[110] OlympIOs, "Integrated optics software," "Available at http://www.c2v.nl/fr_index.shtml?/products/software/olympios-software.shtml".

BIBLIOGRAPHY

[111] X. Li, T. Abe, and M. Esashi, "Deep reactive ion etching of Pyrex glass using SF6 plasma," Sens. Actuators, A **87**, 139–145 (2001).

[112] D. A. Zeze, R. D. Forrest, J. D. Carey, D. C. Cox, I. D. Robertson, B. L. Weiss, and S. R. P. Silva, "Reactive ion etching of quartz and Pyrex for microelectronic applications," J. Appl. Phys. **92**, 3624–3629 (2002).

[113] L. Li, T. Abe, and M. Esashi, "Smooth surface glass etching by deep reactive ion etching with SF6 and Xe gases," J. Vac. Sci. Technol. B **21**, 2545–2549 (2003).

[114] H. C. Jung, W. Lu, S. Wang, L. J. Lee, and X. Hu, "Etching of Pyrex glass substrates by inductively coupled plasma reactive ion etching for micronanofluidic devices," J. Vac. Sci. Technol. B **24**, 3162–3164 (2006).

[115] C. Hunziker, S.-J. Kwon, H. Figi, F. Juvalta, O.-P. Kwon, M. Jazbinšek, and P. Günter, "Configurationally locked, phenolic polyene organic crystal 2-{3-(4-hydroxystyryl)-5,5-dimethylcyclohex-2-enylidene}malononitrile: Linear and nonlinear optical properties," J. Opt. Soc. Am. B **25**, 1678–1683 (2008).

[116] H. Figi, L. Mutter, C. Hunziker, M. Jazbinšek, P. Günter, and B. J. Coe, "Extremely large nonresonant second-order nonlinear optical response in crystals of the stilbazolium salt DAPSH," J. Opt. Soc. Am. B **25**, 1786–1793 (2008).

[117] M. Gad, D. Yevick, and P. E. Jessop, "High-speed polymer/silicon on insulator ring resonator switch," Opt. Eng. **47**, 094601 (2008).

[118] J. H. Wülbern, M. Schmidt, U. Hübner, R. Boucher, W. Volksen, Y. Lu, R. Zentel, and M. Eich, "Polymer based tuneable photonic crystals," phys. stat. sol. (a) **204**, 3739–3753 (2007).

[119] M. Balakrishnan, M. Faccini, M. B. J. Diemeer, E. J. Klein, G. Sengo, A. Driessen, W. Verboom, and D. N. Reinhoudt, "Microring resonator based modulator made by direct photodefinition of an electro-optic polymer," Appl. Phys. Lett. **92**, 153310 (2008).

[120] C. Hunziker, S.-J. Kwon, H. Figi, M. Jazbinšek, and P. Günter, "Fabrication and phase modulation in organic single-crystalline configurationally locked, phenolic polyene OH1 waveguides," Opt. Express **16**, 15903–15914 (2008).

[121] H. Figi, M. Jazbinšek, C. Hunziker, M. Koechlin, and P. Günter, "Electro-optic single-crystalline organic waveguides and nanowires grown from the melt," Opt. Express **16**, 11310–11327 (2008).

[122] P. Günter, C. Bosshard, K. Sutter, H. Arend, G. Chapuis, R. J. Twieg, and D. Dobrowolski, "2-cyclooctylamino-5-nitropyridine, a new nonlinear optical crystal with orthorhombic symmetry," Appl. Phys. Lett. **50**, 486–488 (1987).

BIBLIOGRAPHY

[123] C. Bosshard, K. Sutter, and P. Günter, "Linear- and nonlinear-optical properties of 2-cyclooctylamino-5-nitropyridine," J. Opt. Soc. Am. B **6**, 721–725 (1989).

[124] C. Bosshard, K. Sutter, R. Schlesser, and P. Günter, "Electro-optic effects in molecular crystals," J. Opt. Soc. Am. B **10**, 867–885 (1993).

[125] A. Leyderman, M. Espinosa, T. V. Timofeeva, R. D. Clark, D. O. Frazier, and B. G. Penn, "Growth and characterization of crystalline films of meta-nitroaniline (mNA) and 2-cyclo-octylamino-5-nitropyrydine (COANP)," Proc. SPIE **2809**, 144–154 (1996).

[126] A. Yariv, "Universal relations for coupling of optical power between microresonators and dielectric waveguides," Electron. Lett. **36**, 321–322 (2000).

[127] SCHOTT AG, "Dielektrische Konstante," (2009). "Available at http://www.schott.com".

[128] Q. Xu, D. Fattal, and R. G. Beausoleil, "Silicon microring resonators with 1.5 µm radius," Opt. Express **16**, 4309–4315 (2008).

[129] R. M. Briggs, M. Shearn, A. Scherer, and H. A. Atwater, "Wafer-bonded single-crystal silicon slot waveguides and ring resonators," Appl. Phys. Lett. **94**, 021106 (2009).

[130] W. N. Herman and L. M. Hayden, "Maker fringes revisited: Second-harmonic generation from birefringent or absorbing materials," J. Opt. Soc. Am. B **12**, 416–427 (1995).

[131] B. Wyncke and F. Brehat, "Calculation of the effective second-order non-linear coefficients along the phase matching directions in acentric orthorhombic biaxial crystals," J. Phys. B **22**, 363–376 (1989).

[132] D. W. Berreman, "Optics in stratified and anisotropic media: 4x4-matrix formulation," J. Opt. Soc. Am. **62**, 502–510 (1972).

[133] W. L. Bond, "Measurement of the refractive indices of several crystals," J. Appl. Phys. **36**, 1674–1677 (1965).

[134] S. M. Marazita, W. L. Bishop, J. L. Hesler, K. Hui, W. E. Bowen, and T. W. Crowe, "Integrated GaAs Schottky mixers by spin-on-dielectric wafer bonding," IEEE Transactions on Electron Devices **47**, 1152–1157 (2000).

BIBLIOGRAPHY

Acknowledgments

The realization of the exciting projects of my PhD wouldn't have been possible without all people I had the pleasure to work and spend time with for the last three years. I would like to thank the whole Nonlinear Optics Group for contributing to a great working atmosphere during my PhD. I wish to acknowledge the help of various people who have contributed to the completion of this work:

- Prof. Peter Günter for giving me the possibility to make my PhD in his group, for keeping faith with me and also believing in the final results since the beginning of my work.

- Prof. Markus Sigrist for accepting to be the co-examiner of this thesis.

- Mojca Jazbinšek for all the corrections and discussions and for teaching me the various skills of academic research. I could not have imagined having a better team leader for my PhD.

- Christoph Hunziker, with whom I have shared from the very beginning the same office, for many helpful scientific discussions, new ideas, and for his help by plenty of Matlab and lab electronic problems.

- Manuel Koechlin for all the wafer sawing and writing of photolithography masks with his home-built laser lithography setup.

- Marcel Sturzenegger for technical help and for sharing the same office during the last months of my stay in the group.

- Hansruedi Scherrer and his team for the accomplishment of all e-gun deposition processes and many helpful technical suggestions.

- Christian Herzog, Paolo Losio, Fabian Brunner, and Pepino Sulser for the support with computer problems and the great IT infrastructure of the NLO group.

Acknowledgments

- Seong-Ji Kwon and O-Pil Kwon for the synthesis of plenty of new compounds and the growth of the crystals.

- Jaroslav Hajfler for his expert sample preparation and polishing.

- All members of the NLO group for many entertaining lunch-time discussions and good times together.

- My parents for a lifetime of support and their constant interest in my work.

- Stefanie Croci-Maspoli, my wife, for her love and patience during the PhD period. Her continuing support, encouragement and friendship has been an invaluable resource of strength and advice.

Die VDM Verlagsservicegesellschaft sucht für wissenschaftliche Verlage abgeschlossene und herausragende

Dissertationen, Habilitationen, Diplomarbeiten, Master Theses, Magisterarbeiten usw.

für die kostenlose Publikation als Fachbuch.

Sie verfügen über eine Arbeit, die hohen inhaltlichen und formalen Ansprüchen genügt, und haben Interesse an einer honorarvergüteten Publikation?

Dann senden Sie bitte erste Informationen über sich und Ihre Arbeit per Email an *info@vdm-vsg.de*.

Sie erhalten kurzfristig unser Feedback!

VDM Verlagsservicegesellschaft mbH
Dudweiler Landstr. 99
D - 66123 Saarbrücken

Telefon +49 681 3720 174
Fax   +49 681 3720 1749

www.vdm-vsg.de

Die VDM Verlagsservicegesellschaft mbH vertritt

Printed by Books on Demand GmbH, Norderstedt / Germany